Flows and Chemical Reactions in Homogeneous Mixtures

Flows and Chemical Reactions in Homogeneous Mixtures

Roger Prud'homme

iSTE

WILEY

First published 2013 in Great Britain and the United States by ISTE Ltd and John Wiley & Sons, Inc.

ISTE Ltd
27-37 St George's Road
London SW19 4EU
UK

www.iste.co.uk

John Wiley & Sons, Inc.
111 River Street
Hoboken, NJ 07030
USA

www.wiley.com

Library of Congress Control Number: 2013945046

British Library Cataloguing-in-Publication Data
A CIP record for this book is available from the British Library
ISBN: 978-1-84821-633-4

MIX
Paper from responsible sources
FSC
www.fsc.org
FSC® C013604

Printed and bound in Great Britain by CPI Group (UK) Ltd., Croydon, Surrey CR0 4YY

Table of Contents

List of Main Symbols

Latin characters

a	surface strain rate
a,b	partial derivatives of pressure with respect to ξ and A respectively
A	chemical affinity; chemical species; or monatomic species
A_2	diatomic species
$\underline{\mathrm{A}}$	column matrix of chemical affinities in a multi-reactive medium
A,B	Arrhenius coefficients
c	speed of sound; or molecular speed
c_K	characteristic celerity in a multi-reactive mixture
$c_1, c_\mu, c_\varepsilon, c_k$	coefficients of the $k-\varepsilon$ method
C	total number of moles per unit volume; or Germano's parameter
C_f	friction coefficient
C_j	molar concentration per unit volume

C_p, C_v	specific heat at constant pressure or constant volume respectively (c_p, c_v for the unit of mass)
Cr	crispation number
d	molecular diameter; distance; or differential
D	diffusion coefficient; or diameter
$\vec{\vec{\mathrm{D}}}$	strain rate tensor
Da	Damköhler number
D_T	thermal diffusion coefficient
D_T, D_p	partial derivatives of $\ln n$ with respect to $\ln T$ and $\ln p$ respectively
e	roughness thickness
$\vec{\mathrm{e}}_i$	orthonormal basis vector
E	internal energy (e per unit mass)
$E(t_e)$	residence time distribution in a chemical reactor
$E(k)$	energy spectrum of turbulence
E_a	activation energy
$\mathcal{E}, \mathcal{E}_j$	chemical species
f	parameter; or reduced chemical production rate
f', f''	Reynolds, Favre fluctuation, respectively
$\vec{f}$	force acting on each unit mass
$\vec{f}_j$	force acting on the unitary mass of the species j

F	Helmholtz free energy (f for the unit of mass); generalized force; or any extensive value (f for the unit of mass)
$\vec{\mathrm{F}}$	force
G	Gibbs free enthalpy (g for the unit of mass); or production rate of entities per unit volume of the phase space
$G(\vec{\mathrm{x}},t)$	filter in the physical space
$G(\vec{\mathrm{k}},\omega)$	filter in the Fourier space
$\vec{\mathrm{g}}$	acceleration due to gravity (of modulus g)
g_j	chemical potential per unit mass of the species j in a mixture
H	enthalpy (h per unit mass)
h_0	stagnation enthalpy
$\vec{\vec{\mathrm{I}}}$	unit tensor
$I(\alpha)$	distribution of ages
j	chemical species
$\vec{\jmath}_{Dj}$	diffusion flux of the species j
J_0, J_1	Bessel functions
k	Boltzmann's constant; wave number; or kinetic energy of turbulence
$k(T)$	specific reaction rate
K	kinetic energy (k per unit mass); compressibility; number of chemical reactions in a mixture; heat exchange coefficient; or wave number

K_F	turbulent exchange coefficient for the quantity F
K_C, K_p	equilibrium constants for the concentrations and the partial pressures, respectively
l	latent heat per unit mass; or mean free path
ℓ	length of transfer; or integral scale of turbulence
L	length; molar latent heat; number of chemical elements in a mixture; size of a crystal; or phenomenological coefficient of a chemical reaction
ℓ_D	diffusion thickness of a non premixed flame
ℓ_f, ℓ_δ	thickness of preheating, of reaction of a premixed flame, respectively ℓ_G, ℓ_K: length scales of Gibson, of Kolmogorov, respectively
Le	Lewis number
Lp	Prandtl mixing length
$\underline{\underline{L}}$	matrix of phenomenological coefficients of chemical reactions
$\mathcal{L}$	resolved strained tensor (in LES)
$\mathcal{L}(f)$	derivation operator in cylindrical coordinates
m	total mass
M	molecular mass; diluent; or Mach number
$\mathcal{M}$	molar mass
m_j	mass of the species j
$\mathcal{M}_j$	molar mass of the species j
$\dot{m}$	unit mass flow rate; or mass flow rate of a nozzle

n	total number of moles
n_j	number of moles of the species j
N	number of species; number of molecules per unit volume; or coordinate normal to an interface
$\vec{n}, \vec{N}$	unitary normal to an interface
p	thermodynamic pressure
P	probability density
$\vec{\vec{P}}$	pressure tensor
Pr	Prandtl number
q	any parameter; or heat flux
$\dot{q}$	volume flowrate
$\vec{q}$	heat flux vector
Q	partition function; or quantity of heat
$\dot{Q}$	heat released per unit time at the walls of a chemical reactor
$\left(Q_f^0\right)_j = \left(H_0^0\right)_j$	molar enthalpy of formation of the species j ($\left(q_f^0\right)_j$ per unit mass)
r	constant of perfect gases per unit mass; radius; or caloric power received per unit volume
R	universal molar gas constant; radius; or number of independent chemical species in a mixture
$R(\xi)$	correlation coefficient
Re	Reynolds number

R_j	mass production of species j by chemical reaction
s	Arrhenius exponent
S	entropy (s per unit mass); area of the cross-section of a nozzle
s_L, s_L^0, s_t	laminar combustion velocity, standard, and turbulent, respectively
S, $\mathcal{S}$	surface
S	symmetrical part of the velocity gradient tensor
Sc	Schmidt number
t	time; diffusion thickness
t_e	residence time in a chemical reactor
T	absolute temperature; or Chapmann-Jouguet detonation point
T_a, T_{ad}	activation temperature, adiabatic temperature of a reaction, respectively
$\vec{\vec{\mathrm{T}}}, \mathrm{T}$	double-filter tensor (in LES)
$\vec{\vec{\mathcal{T}}}, \mathcal{T}$	unresolved Reynolds tensor (in LES)
u, v, w	velocity $\vec{\mathrm{v}}$ components in Cartesian coordinates (v_r, v_θ, v_z in cylindrical coordinates)
U, U_∞	reference velocity
$\vec{\mathrm{U}}, \vec{\mathrm{v}}$	velocity vector; or barycentric velocity vector in a composite fluid
v, v'	speed, turbulence intensity respectively
V	speed; force; volume in the phase space; or potential

$\vec{V}$	vector; velocity vector; or velocity vector in the phase space $(\vec{x},\vec{\zeta})$
$\mathcal{V}$	volume; or control volume
$\vec{v}_j$	velocity vector of the species j
$\vec{V}_j$	diffusion velocity of the species $\vec{v}_j - v$
$\vec{w}$	velocity of a surface (normal component w); or $d\vec{\zeta}/dt$ in the phase space
$\vec{W}$	local velocity vector of a discontinuity
$\dot{W}_F$	rate of production of the quantity F ($\dot{W}_j$ for species j)
$\dot{W}_{E\alpha,\text{int}}$	rate of production of energy for the internal degrees of freedom of the species j
x, y, z	Cartesian coordinates; x along a nozzle axis
$\vec{x}$	position vector
X_j, Y_j	molar and mass fraction of the species j respectively
Z	fraction of mixture

Greek symbols

α	species; or age of an entity in a chemical reactor
β_j, β_T	reduced concentration, respectively reduced temperature
δ	thickness of a viscous layer; $\delta(x)$: Dirac distribution
Δ	difference; Laplacian; size of a filter (in LES); ΔH : heat of a reaction

ε	small dimensionless parameter; or turbulent dissipation rate
ε_δ	relative roughness
φ	velocity potential; $\varphi(\vec{x},t)$: weight function
ϕ_j	partial molar quantity associated with the quantity ϕ
γ	isentropic coefficient c_p/c_v ; damping of a wave
Γ	circulation of a vortex; second partial derivative of enthalpy
χ	scalar local dissipation rate
η	partial bulk viscosity; or reduced coordinate
κ	thermal diffusivity $\lambda/\rho c_p$; or mean curvature of a surface
λ	coefficient of thermal conductivity; eigenvalue; Taylor's micro-scale
Λ	coefficient of head-loss; or heat transfer coefficient
μ	coefficient of shear viscosity; Gibbs free energy per mole; or absorption coefficient per unit of wave length
μ_j	molar chemical potential of a species j in a mixture
ν	kinematic viscosity μ/ρ
ν_t	turbulent kinematic viscosity
ν_j	algebraic stoichiometric coefficient $\nu_j = \nu''_j - \nu'_j$
ν'_j, ν''_j	stoichiometric coefficient of the direct reaction, or its inverse respectively
$\bar{\bar{\Pi}}$	viscous pressure tensor

Π_i	dimensionless group
θ	temperature; or angular coordinate
ϑ	volume per unit mass (inverse of the density)
ρ	density (volumetric mass)
ρ_j	partial density
σ	surface tension
Σ	surface; area of a surface; $\Sigma(x)$ area of the cross-section of a nozzle
$\vec{\vec{\Sigma}}$	stress tensor
$\sigma_k, \sigma_\varepsilon$	Prandtl numbers of the $k-\varepsilon$ method
τ	characteristic time; crossing time in a chemical reactor; dimensionless energy of reaction $\tau = \Delta H / c_p T_1$
ω	speed of rotation; or pulsation of an oscillating wave; function of β_T
$\vec{\omega}$	rotation vector
Ω	speed of rotation
$\vec{\Omega}$	rotation vector
ξ	progress variable per unit mass; reduced coordinate; or correlation length
$\underline{\xi}$	column matrix of the progress variables
ψ	stream function; or probability in the space phase
ζ	progress variable per unit volume; or reduced variable

$\dot{\zeta}$ rate of production of a chemical reaction

$\vec{\zeta}$ vector of the phase space

Subscripts, superscripts, and other symbols

a of activation; or relative to the quantities per unit area of the interface

ad adiabatic

b burned gases

c concentration; or cut-off

C critical point

chem chemical

$CO{-}E_{\beta v}$ coupling CO molecule - vibrational energy

d relative to small dissipative eddies

D direct; of dissociation; or diffusive

e equilibrium flow; exit of a reactor; residence; or large eddies

eff effective

EBU relating to the "Eddy break-up" model

f frozen composition; fresh gases; or flame

G, ^{g} gas

α, β, i, j of species

i internal; relative to imaginary part; or irreversible

int	internal degrees of freedom of a molecule
K	for the K frozen progress variables
l	liquid
L	line; liquid; or laminar
m	mixture; mass
mec	mechanical
p	at constant pressure; or solid phase
q	extinction limit of a flame
r	chemical reaction; or reference
R	reverse; or recombination
$R-K$	relative to the $R-K$ progress variables at equilibrium
s	steady state; surface; or isentropic
o	entry of a reactor
S	surface; relative to the specific or intensive interfacial quantities
st	stoichiometric; or steady
t	for translational energy mode of a molecule; turbulent
T	temperature; turbulent; or at constant temperature
T	second order tensor; or transpose of a tensor
$\circ$	deviator of a tensor
th	thermal

$_v$, or $_\vartheta$	at constant volume
V	vapor
$//$	parallel to a surface
$\perp$	normal to a surface
0	standard reference value
$\bullet$	pure simple substance
$\dot{}$	per unit time; or for a rate of production
$\bar{}$	thermodynamic value per mole; average quantity; or Reynolds average
$'$	Reynolds disturbance in relation to an average value
$''$	Favre disturbance in relation to an average value
$S(\ )$	symmetrical part of a matrix or a tensor
$\tilde{}$	transposed tensor; transposed matrix; or Favre average
$\langle\ \rangle$	ensemble average
$\hat{}$	pre-exponential factor; or relative to a test filter in LES
$(\ )_T^0$	standard thermodynamic function
$\times$	vector product
$\otimes$	tensor product
$\cdot$	scalar product (singly-contracted tensor product)
$:$	dyadic product (doubly-contracted tensor product)
$\wedge$	exterior product

*	sonic conditions; or reference state; or virtual
$\vec{\nabla}$	nabla (gradient operator)
$[\ ]_{-}^{+}$	jump of a quantity across an interface
d/dt	material derivative equal to $\partial/\partial t + \vec{v} \cdot \vec{\nabla}$
$\partial/\partial t$	partial time derivative
$d_{\mathbf{W}}/dt$	material derivative associated with the velocity $\vec{\mathbf{W}}$ equal to $\partial/\partial t + \vec{\mathbf{W}} \cdot \vec{\nabla}$
$\rightleftarrows$	for a reversible chemical reaction

Preface

This book – a follow-up to the author's previous publication, *Flows and Chemical Reactions* [PRU 12] – is devoted to the applications of such flows and reactions.

Remember that the former book on *general equations* comprised three chapters:

1. Simple fluids;

2. Reactive mixtures; and

3. Interfaces and lines.

It is unsurprising that the examples chosen for this book relate mainly to reactive homogeneous mixtures, because on the one hand, numerous examples of non-reactive fluids were discussed in Chapter 1 of the earlier work and, on the other, heterogeneous flows, notably including interfaces and lines, will be the topic of a third, forthcoming publication.

Diffusion phenomena are overlooked in Chapter 1, but the reactions are of crucial importance. Thus, the focus is on the propagation of sound and on mono-dimensional flows in nozzles. Damköhler numbers and differences from chemical equilibrium play an important part, and relate to the performances of thrusters. In this chapter, the balance equations for non-equilibrium flows are extended to non-equilibrium of the internal modes of molecule energy, which is one way of dealing with the problems of bodies' re-entry into the atmosphere.

Chapter 2, on chemical reactors, is very conventional, and the main point discussed is that of homogeneous, perfectly stirred reactors, where the analytical computations can be pursued further – particularly with regard to the stability of stationary operating points. Many different types of instability are to be found in this chapter, and we can deal with these by using control parameters linked particularly to heat exchanges with the environment. Other types of reactor are also introduced, as are the classic instruments used in experimental and theoretical analysis such as population balance, distribution of residence time and age distribution.

Combustion, discussed in Chapter 3, is an essential domain of application for the concepts learned about in the previous book and the first two chapters of this one, and we examine different cases of it – first in a laminar flow and then in a turbulent flow. However, a distinction is drawn, with regard to flames, between premixed flames and non-premixed flames, which do not arise from the same mechanisms. The chapter also discusses triple flames, which include both aspects at once, and are very useful for those people, such as engineers and researchers, who are specifically interested in autoignition in motors (be they diesel or otherwise).

In the appendices, we recap the logic behind dimensional analysis, which is very widely used to study any fluid flow problem which is even slightly complex. Clarifications are also given with regard to statistical thermodynamics, with the partition functions relating to the different internal modes of the molecules: translation, rotation, electronics, vibration and the connections between these modes and the chemical reactions. Next we present the basic concepts about turbulence: after having examined various sources of destabilization of laminar flows with and without reaction, we turn our attention to pure fluids in incompressible and compressible flows, the RANS and LES methods. Finally, certain specifications are given with regard to bifurcations, flames in a confined medium and diffusion flames.

Roger PRUD'HOMME
August 2013

Chapter 1

Flows in Nozzles

Of the applications for the general equations of flows with chemical reactions [PRU 12], many are specific to combustion, and some of these will be discussed in Chapter 3. In this third chapter, we present a number of problems, dealt with in Chapter 1 in the case of simple fluids, whose solution is altered by the presence of chemical reactions. The propagation of sound, examined in section 1.6.2 of [PRU 12], in the case of an ideal gas, is one of these problems. Compressible flow in a de Laval nozzle (section 1.6.6) is another. In both these problems, the phenomena of dissipation are neglected. Otherwise, we need to take account not only of viscosity and heat conduction but also of the diffusion of species, as seen in particular in section 2.5.

In the present chapter, we continue to neglect the diffusion of species, of the momentum and of energy, both for linearized sound propagation and for flows in nozzles. In sections 1.1 and 1.2, we examine the cases of one or more chemical reactions and, in section 1.3, we add the non-equilibria of the internal modes of energy of the molecules.

1.1. Sound propagation in the presence of chemical reactions

Linear sound propagation occurs differently in a mono-species medium and a multi-species medium with chemical reactions. In the former case, as we saw in section 1.6.2 of [PRU 12], in the absence of dissipative phenomena (viscosity or heat conduction), the celerity of sound had a clearly-defined value at a given temperature and pressure. We shall see in this section that the same is not true in the presence of chemical kinetics, and that the value of the celerity of sound at a given pressure and temperature depends on its frequency and lies between two extreme values called the "frozen" and "equilibrium" values. The former corresponds to a mixture with constant concentrations and is larger than the latter, which for its part corresponds to a mixture at chemical equilibrium. In addition, the soundwave becomes damped, and its attenuation is a function of the frequency.

We shall first examine the case of a single chemical reaction, and then look at the case of multiple reactions.

The results of this study are widely applicable for all problems involving waves, be they steady or otherwise, and more generally anywhere where the Mach number is involved, i.e. in a compressible reactive flow, such as in nozzles (section 1.2) and in combustion waves (Chapter 3).

1.1.1. *Thermodynamic considerations*

In Chapters 1 and 2 of [PRU 12], we gave the meanings of the first derivatives of the internal energy with relation to the extensive values upon which it depends. For a mixture of N chemical species, we obtain

$$E = E\left(S, \mathcal{V}, n_1, n_2 \ldots n_N\right) \quad [1.1]$$

$$T = \partial E/\partial S, -p = \partial E/\partial \mathcal{V}, \mu_j = \partial E/\partial n_j \quad [1.2]$$

The chemical potential g per unit mass was also introduced, with $g_j = \mu_j/\mathcal{M}_j$, which yields results similar to those above, where

$$\begin{cases} E = E\left(S, \mathcal{V}, m_1, m_2 \ldots m_N\right) \\ T = \partial E/\partial S, -p = \partial E/\partial \mathcal{V}, g_j = \partial E/\partial m_j \end{cases} \quad [1.3]$$

Here, we shall further explain what was said in Chapter 2 of [PRU 12] – i.e. that in the case of stable equilibria, the second law of thermodynamics leads to conditions on the second partial derivatives of the internal energy in relation to the extensive variables, and that this is not without consequence in terms of the usual values and relations.

Consider a mixture of N species, distributed on both sides of a non-deformable, adiabatic and impermeable wall of a container which itself is insulated from the outside environment.

Each sub-system is at thermodynamic equilibrium, at fixed concentrations. If we take the wall away (assuming that this requires a negligible amount of work), the mixture occupies the whole of the vessel, and we can see that it reaches a new state of equilibrium, which constitutes a characteristic of stability of that equilibrium. Suppose that no chemical reaction has taken place during the course of the transformation.

We obtain the following relations:

$$\begin{cases} E + E' = E_f, \ S + S' \leq S_f, \\ \mathcal{V} + \mathcal{V}' = \mathcal{V}_f, \ n_j + n'_j = n_{jf} \end{cases} \qquad [1.4]$$

Thus:

$$\begin{aligned} & E(S, V, n_1, n_2 \ldots n_N) + E(S', V', n'_1, n'_2 \ldots n'_N) \\ & = E(S_f, V + V, n_1 + n'_1, n_2 + n'_2 \ldots n_N + n'_N) \end{aligned} \qquad [1.5]$$

Because $\partial E / \partial S = T > 0$:

$$E(S_f, \ldots) \geq E(S + S', \ldots) \qquad [1.6]$$

It follows that

$$E(S, \mathcal{V}, n_j) + E(S', \mathcal{V}', n_j') \geq E(S + S', \mathcal{V} + \mathcal{V}', n_j + n_j') \qquad [1.7]$$

The homogeneous nature of the first degree of the function $E(S, \mathcal{V}, n_j)$ means that we can write

$$\frac{1}{2}\left[E(S, \mathcal{V}, n_j) + E(S', \mathcal{V}', n_j')\right] \geq E\left(\frac{S + S'}{2}, \frac{\mathcal{V} + \mathcal{V}'}{2}, \frac{n_j + n_j'}{2}\right) \qquad [1.8]$$

which leads us to the following conclusion: $E(S, \mathcal{V}, n_j)$ *is a convex function.*

It is easy to demonstrate that this result implies a positive definite matrix of second partial derivatives:

$$\begin{Vmatrix} \partial^2 E/\partial S^2 & \partial^2 E/\partial S\,\partial \mathcal{V} & \partial^2 E/\partial S\,\partial n_1 \ldots & \partial^2 E/\partial S\,\partial n_N \\ \partial^2 E/\partial \mathcal{V}\,\partial S & \partial^2 E/\partial \mathcal{V}^2 & \partial^2 E/\partial \mathcal{V}\,\partial n_1 \ldots & \partial^2 E/\partial \mathcal{V}\,\partial n_N \\ \partial^2 E/\partial n_1\,\partial S & \partial^2 E/\partial n_1\,\partial \mathcal{V} & \partial^2 E/\partial n_1^{\,2} \;\ldots & \partial^2 E/\partial n_1\,\partial n_N \\ \vdots & \vdots & \vdots & \vdots \\ \partial^2 E/\partial n_N\,\partial S & \partial^2 E/\partial n_N\,\partial \mathcal{V} & \partial^2 E/\partial n_N \partial n_1 \;\ldots & \partial^2 E/\partial n_N^{\,2} \end{Vmatrix} \quad [1.9]$$

The necessary and sufficient condition for it to be so, in the case of a symmetrical matrix, is that all the principal minors be positive. Thus:

$$\begin{cases} \partial^2 E/\partial S^2 = \partial T/\partial S > 0, \partial^2 E/\partial \mathcal{V}^2 = -\partial p/\partial \mathcal{V} > 0 \\ \partial^2 E/\partial S^2\, \partial^2 E/\partial \mathcal{V}^2 - (\partial^2 E/\partial S\,\partial \mathcal{V})^2 > 0 \\ \partial^2 E/\partial n_j^{\,2} = \partial \mu_j/\partial n_j > 0, \text{ etc.} \end{cases} \quad [1.10]$$

Similar results are obtained by replacing the numbers of moles n_j with the masses m_j of the species.

What are the consequences of doing this for the unit of mass of the mixture? We need to take account of the relation:

$$\sum_{j=1}^{N} \mathcal{M}_j n_j = 1 \quad [1.11]$$

This additional relation does not alter the convexity of the function E which then becomes e, and therefore depends on N+1 independent variables instead of N+2.

This result can be generalized for cases where we also impose a dependency of the concentrations of a certain number R of independent progress variables ξ_r:

$$n_j = n_j^0 + \sum_{r=1}^{R} \nu_{jr}\, \xi_r \quad [1.12]$$

where

$$\sum_{j=1}^{N} \mathcal{M}_j n_j^0 = 1 \quad [1.13]$$

and

$$\sum_{j=1}^{N} \nu_{jr} \mathcal{M}_j = 0 \qquad [1.14]$$

With A_r representing the corresponding chemical affinities, we obtain

$$A_r = -\sum_{j=1}^{N} \nu_{jr} \mu_j \qquad [1.15]$$

Thus

$$\begin{cases} e = e(s, \vartheta, \xi_1, \xi_2, \ldots \xi_R) \\ de = T\,ds - p\,d\vartheta - \sum_{r=1}^{R} A_r\,d\xi_r \end{cases} \qquad [1.16]$$

By using the notation $\underline{\mathbf{A}}$ to represent the column matrix of the chemical affinities with the A_r values as components, and $\underline{\xi}$ to represent the column matrix of the independent progress variables ξ_r, the Gibbs equation [1.16] can also be written as

$$\begin{cases} e = e(s, \vartheta, \underline{\xi}) \\ de = T\,ds - p\,d\vartheta - \tilde{\underline{\mathbf{A}}}\,d\underline{\xi} \end{cases}$$

Using a classic convention, let us set

$$\partial^2 e / \partial x\, \partial y = e_{xy} \qquad [1.17]$$

The convexity of the internal energy function results in the matrix being positive definite:

$$\begin{Vmatrix} e_{ss} & e_{s\vartheta} & e_{s\xi_1} & \cdots & e_{s\xi_R} \\ e_{\vartheta s} & e_{\vartheta\vartheta} & e_{\vartheta\xi_1} & \cdots & e_{\vartheta\xi_R} \\ e_{\xi_1 s} & e_{\xi_1\vartheta} & e_{\xi_1\xi_1} & \cdots & e_{\xi_1\xi_R} \\ \vdots & \vdots & \vdots & \cdots & \vdots \\ e_{\xi_R s} & e_{\xi_R\vartheta} & e_{\xi_R\xi_1} & \cdots & e_{\xi_R\xi_R} \end{Vmatrix} \qquad [1.18]$$

with the following sub-matrices:

$$\mathbf{e}_{\xi s} = -\underline{\mathbf{A}}_s = \left\| \begin{matrix} e_{\xi_1 s} \\ \vdots \\ e_{\xi_R s} \end{matrix} \right\|, \mathbf{e}_{\xi \vartheta} = -\underline{\mathbf{A}}_\vartheta = \left\| \begin{matrix} e_{\xi_1 \vartheta} \\ \vdots \\ e_{\xi_R \vartheta} \end{matrix} \right\|, \mathbf{e}_{\xi\xi} = -\underline{\mathbf{A}}_\xi = \left\| \begin{matrix} e_{\xi_1\xi_1} & \cdots & e_{\xi_1\xi_R} \\ \vdots & \cdots & \vdots \\ e_{\xi_R\xi_1} & \cdots & e_{\xi_R\xi_R} \end{matrix} \right\|$$

Let us look at the consequences of this.

In the case of a simple fluid (i.e. a fluid with only one component), we have the inequalities [1.22], and can therefore define the specific heats:

$$c_v = T\left(\partial s/\partial T\right)_\vartheta = T/e_{ss} > 0 \text{ and}$$
$$c_p = T\left(\partial s/\partial T\right)_p = T/\left(e_{ss} - e_{s\vartheta}{}^2/e_{\vartheta\vartheta}\right) > 0$$

Also, $c_p - c_v = T/\left[e_{ss}\left(e_{ss}e_{\vartheta\vartheta}/e_{s\vartheta}{}^2 - 1\right)\right] > 0$. We can also define the characteristic celerity c using the equation:

$$c^2 = \left(\partial p/\partial \rho\right)_s = -\vartheta^2\left(\partial p/\partial \vartheta\right)_s = \vartheta^2 e_{\vartheta\vartheta} > 0$$

In the case of a mixture with one chemical DOF, and if the mixture has a fixed chemical composition, the results are similar to those shown above:

$$\begin{cases} c_{vf} > 0, c_{pf>0}, c_{pf} - c_{vf} > 0, \\ c_f{}^2 = \left(\partial p/\partial \rho\right)_{s,\xi} \end{cases} \qquad [1.19]$$

If the mixture is at chemical equilibrium, we have

$$\begin{cases} A = 0, c_{ve} = T\left(\partial s/\partial T\right)_{\vartheta,A} = T/\left(e_{ss} - e_{s\xi^2}/e_{\xi\xi}\right) > 0, \\ c_{pe>0}, c_{pe} - c_{ve} > 0, \\ c_e{}^2 = \left(\partial p/\partial \rho\right)_{s,A} = \vartheta^2\left(e_{\vartheta\vartheta} - e_{\vartheta\xi}{}^2/e_{\xi\xi}\right) \le c_f{}^2 \end{cases} \qquad [1.20]$$

However, we know nothing, at this point, about the signs of quantities such that:

$$a = \left(\partial p/\partial \xi\right)_{s,\rho} = -e_{\vartheta\xi} \qquad [1.21]$$

$$b = (\partial p/\partial A)_{s,\rho} = e_{\vartheta\xi}/e_{\xi\xi} \quad [1.22]$$

The ratio of *b/a*, equal to $(-1/e_{\xi\xi})$ is for its part always negative.

In the case of a mixture with several chemical DOFs:

– the inequalities obtained are the same as for a single DOF;

– we can envisage thermodynamic states for which some of the ξ_r are constant and the A_r values corresponding to the other ξ_r are zero. We then define new inequalities, specific heats and characteristic celerities such that: $c_K{}^2 = (\partial p/\partial \rho)_{s,\xi_1,\dots\xi_K,A_{K+1},\dots A_R}$.

1.1.2. *Sound propagation in a mono-reactive medium*

Sound propagation in a medium with one chemical DOF was studied by Einstein [EIN 20] and more recently, notably, by Wegener *et al.* [WEG 65].

We envisage a chemical reaction which is taken to be reversible and near to equilibrium. If ξ represents its progress variable, in the absence of diffusion, we have

$$\rho\, d\xi/dt = L\,A \quad [1.23]$$

where A is the chemical affinity and L the corresponding phenomenological coefficient (see Chapter 2 of [PRU 12]).

The other equations involved in the problem are

$$\begin{cases} d\rho/dt + \rho\vec{\nabla}\cdot\vec{\mathbf{v}} = 0 \\ \rho\, d\,\vec{\mathbf{v}}/dt + \vec{\nabla}p = \vec{\mathbf{0}} \\ \rho\, d\,s/dt = L\,A^2/T \end{cases} \quad [1.24]$$

To this system, we have to add the equations of state of the mixture or its fundamental energy law, e.g. in the form

$$e = e(s,\vartheta,\xi),\ \vartheta = 1/\rho \quad [1.25]$$

The reference state is characterized by

$$\rho = \rho_0, p = p_0, \vec{\mathbf{v}} = \vec{\mathbf{0}}, A = 0, \xi = \xi_0 \qquad [1.26]$$

and the slightly disturbed state by

$$\begin{cases} \rho = \rho_0 + \rho', p = p_0 + p', \vec{\mathbf{v}} = \vec{\mathbf{v}}' \\ s = s_0 + s', A = A', \xi = \xi_0 + \xi' \end{cases} \qquad [1.27]$$

Linearized theory then gives us:

$$\rho_0 \, \partial \xi' / \partial t = L_0 A' \qquad [1.28]$$

$$\partial \rho' / \partial t + \rho_0 \vec{\nabla} \cdot \vec{\mathbf{v}}' = 0 \qquad [1.29]$$

$$\rho_0 \, \partial \vec{\mathbf{v}}' / \partial t + \vec{\nabla} p' = \vec{\mathbf{0}} \qquad [1.30]$$

$$s' = 0 \qquad [1.31]$$

If we choose s, p and ξ to be the thermodynamic variables, we have:

$$p' = c_{f0}{}^2 \rho' + a_0 \xi' \qquad [1.32]$$

where $c_f = \left(\partial p / \partial \rho \right)_{s,\xi}{}^{1/2}$ is the local speed of sound of the medium, whose composition would be fixed. The celerity c_f is identical to the celerity c of a non-reactive fluid. In addition, $a = \left(\partial p / \partial \xi \right)_{s,\rho}$ is a thermodynamic quantity (equation [1.21]).

If we now take s, p and A as variables, we have

$$p' = c_{e0}{}^2 \rho' + b_0 A' \qquad [1.33]$$

with $c_e = \left(\partial p / \partial \rho \right)_{s,A}{}^{1/2}$ where c_e is the speed of sound in a mixture at chemical equilibrium (this is defined by $A = 0$), and where $b = \left(\partial p / \partial A \right)_{s,\rho}$ (equation [1.22]).

NOTE.– We know that c_f and c_e are positive real numbers, we do not *a priori* know the signs of a and b, but we do know that the ratio b/a is negative (equations [1.21] and [1.22]). If we envisage a chemical reaction at constant volume, in a near-equilibrium state we have

$$\rho_0\,\partial\xi'/\partial t = L_0\,A', p' = a_0\,\xi' \qquad [1.34]$$

so that

$$\partial\xi'/\partial t - \left(L_0\,a_0/\rho_0\,b_0\right)\xi' = 0 \qquad [1.35]$$

and $\tau_{v0} = -\rho_0 b_0 / L_0 a_0 > 0$ represents the characteristic time of the reaction.

Similarly, with a reaction at constant pressure, we would have

$$\tau_{p0} = -\rho_0 b_0 c_{f0} / L_0 a_0 c_{e0} > 0$$

From equation [1.30], we can deduce that

$$\vec{\mathbf{v}}' = \vec{\nabla}\varphi', p' = -\rho_0\,\partial\varphi'/\partial t \qquad [1.36]$$

The continuity equation [1.29] gives us

$$c_{f0}^{\ -2}\left(\partial p'/\partial t - a_0\,\partial\xi'/\partial t\right) + \rho_0 \vec{\nabla}\cdot\vec{\mathbf{v}}' = 0 \qquad [1.37]$$

or, if we use equations [1.28] and [1.37]:

$$c_{f0}^{\ 2}\Delta\varphi' - \partial^2\varphi'/\partial t^2 = \left(L_0 a_0/\rho_0^{\ 2}\right)A' \qquad [1.38]$$

We differentiate both sides of the equation with respect to *t*, and get

$$(\partial/\partial t)\left[c_{f0}^{\ 2}\Delta\varphi' - \partial^2\varphi'/\partial t^2\right] - \left(L_0 a_0/\rho_0^{\ 2} b_0\right)\left[c_{e0}^{\ 2}\Delta\varphi' - \partial^2\varphi'/\partial t^2\right] = 0 \qquad [1.39]$$

This equation, which governs the motion of small disturbances, can, in light of the above, be written in one of the following two equivalent forms:

$$\begin{cases} \tau_{v0}\,(\partial/\partial t)\left[\partial^2\varphi'/\partial t^2 - c_{f0}^{\ 2}\Delta\varphi'\right] + \partial^2\varphi'/\partial t^2 - c_{e0}^{\ 2}\Delta\varphi' = 0 \\ \tau_{p0}\,(\partial/\partial t)\left[c_{f0}^{\ -2}\partial^2\varphi'/\partial t^2 - \Delta\varphi'\right] + c_{e0}^{\ -2}\partial^2\varphi'/\partial t^2 - \Delta\varphi' = 0 \end{cases} \qquad [1.40]$$

The partial differential equation in φ' is linear. We can look for elementary solutions of the type:

$$\varphi' = \hat{\varphi} e^{i\left(\vec{\mathbf{K}}\cdot\vec{\mathbf{x}}-\omega t\right)} \quad [1.41]$$

representing free plane waves with frequency $\omega/2\pi$. For $\vec{\mathbf{K}}$ and ω, we obtain the following equation, giving K^2 as a function of ω:

$$K^2 = \frac{\omega^2\left(1-i\omega\tau_{v0}\right)}{c_{e0}{}^2 - i\omega\tau_{v0}c_{f0}{}^2} \quad [1.42]$$

Two extreme cases are of interest to us. When $\omega\tau_{v0}$ is close to zero (very slow vibration), the evolution is slow in comparison to the chemical reaction, and chemical equilibrium is maintained at all times, we then find $\omega^2/K^2 \cong c_{e0}{}^2$, and the celerity of propagation of small motions is equal to c_{e0}, which does indeed represent the speed of sound at chemical equilibrium. When $\omega\tau_{v0}$ is large (very rapid vibration), the chemical reactions are frozen, and we find $\omega^2/K^2 \cong c_{f0}{}^2$ as happens in non-reactive media.

In the case of finite values of $\omega\tau_{v0}$, we see the phenomenon of relaxation. The exponent $i\left(\vec{\mathbf{K}}\cdot\vec{\mathbf{x}}-\omega t\right)$ can also be written as $i\omega\left(\vec{\mathbf{K}}\cdot\vec{\mathbf{x}}/\omega - t\right)$. If we decompose the wavenumber vector $\vec{\mathbf{K}}$ into its real part $\vec{\mathbf{K}}_r$ and imaginary part $\vec{\mathbf{K}}_i$, this exponent becomes equal to

$$-\vec{\mathbf{K}}_i\cdot\vec{\mathbf{x}} + i\omega\left(\frac{\vec{\mathbf{K}}_r\cdot\vec{\mathbf{x}}}{\omega} - t\right) \quad [1.43]$$

Thus, K_i characterizes the spatial damping of the wave and K_r/ω is the inverse of its propagation celerity c. The damping of the wave and its celerity are functions of the pulsation ω. These functions cannot easily be determined, except if the characteristic velocities c_f and c_e are similar to one another. In view of equation [1.20], we always have $c_f \geq c_e$. The condition is therefore written thus:

$$c_{f0}{}^2 = c_{e0}{}^2\left(1+\varepsilon\right) \quad [1.44]$$

If we set

$$K = K_r + iK_i, K_r/\omega = 1/c(\omega), K_i/\omega = \gamma(\omega) \quad [1.45]$$

we obtain the following, after linearization in ε and separation of the real and imaginary parts:

$$\begin{cases} 1/c^2 - \gamma^2 = c_{e0}^{-2}\left[1 - \varepsilon\omega^2\tau_{v0}^2/\left(1+\omega^2\tau_{v0}^2\right)\right] \\ 2\gamma/c = \left[\left(c_{f0} - c_{e0}\right)/c_{e0}^2\right]\left[\omega\tau_{v0}/\left(1+\omega^2\tau_{v0}^2\right)\right] \end{cases} \quad [1.46]$$

Solving this system, whilst neglecting the terms in ε^2, gives us

$$\begin{cases} c^2(\omega) = \left(c_{e0}^2 + \omega^2\tau_{v0}^2 c_{f0}^2\right)/\left(1+\omega^2\tau_{v0}^2\right) \\ \gamma = \left[\left(c_{f0} - c_{e0}\right)/c_{e0}^2\right]\left[\omega\tau_{v0}/\left(1+\omega^2\tau_{v0}^2\right)\right] \end{cases} \quad [1.47]$$

The attenuation of the wave per unit wavelength is

$$2\pi c\gamma = 2\pi\left[\left(c_{f0} - c_{e0}\right)/c_{e0}\right]\left[\omega\tau_{v0}/\left(1+\omega^2\tau_{v0}^2\right)\right] \quad [1.48]$$

The expression of $c(\omega)$ is compatible with the limits found for a small $\omega\tau_{v0}$ value and an infinite $\omega\tau_{v0}$ value. The curves in Figure 1.1 illustrate these results.

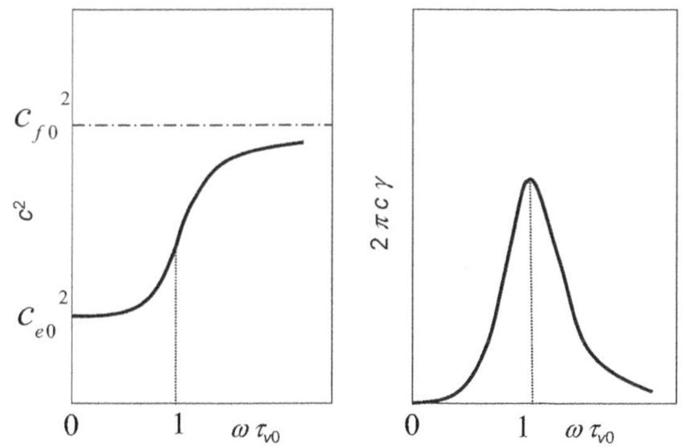

Figure 1.1. *Celerity and attenuation of the soundwave propagating in a mono-reactive gaseous mixture*

In conclusion, the small motions of a reactive fluid propagate at a velocity which depends on the frequency of the monochromatic plane wave envisaged, and they attenuate in space, with the coefficient of attenuation per unit wavelength also being a function of the frequency, in accordance with the laws discussed above.

Let us now apply this result to a piston animated with small motions around the position $x = 0$, with frequency $\omega/2\pi$. The wave amplitude varies with $e^{-K_i x}$ and, if $\left(c_{f0}^2 - c_{e0}^2\right)/c_{e0}^2 << 1$, then K_i is equal to

$$K_i = \frac{\left(c_{f0} - c_{e0}\right)\omega^2\tau_{v0}}{c_{e0}^2\left(1 + \omega^2\tau_{v0}^2\right)}$$

For a given frequency, the attenuation depends on τ_{v0}; it is maximum for $\tau_{v0} = 1/\omega$, at a given abscissa value. It is proportional to x, the distance from the origin.

Note that $K_i = 0$ for $\omega = 0$ (equilibrium) and that $K_i \to \varepsilon/2c_{e0}\tau_{v0}$ as $\omega \to \infty$ (freezing). The low-frequency wave is therefore not greatly attenuated, as it is the slowest (velocity c_{e0}), whereas the high-frequency wave is the most attenuated at a fixed τ_{v0}. In the presence of small non-monochromatic motions of the piston, there is a combination of these waves with an amplitude distribution depending on the frequency.

1.1.3. *Sound propagation in a multi-reactive medium*

1.1.3.1. *General points*

Let us look at the propagation of an unsteady small disturbance in a multi-reactive medium at rest and at chemical equilibrium in the reference state. Many authors have examined this problem in the case of multiple chemical DOFs [MEI 59; BAU 65; KNE 65; NAP 64; PRU 68].

Here, we give a simple method for determining the different characteristic times involved in the dispersion relation [BAR 73; PRU 68].

We make the hypothesis of a fluid or homogeneous mixture of fluids, with the same properties in all directions.

Consider the disturbance to be a sinusoidal function of time at all points in the medium, in accordance with equation [1.41], but this hypothesis is not restrictive because we can decompose a non-sinusoidal disturbance into a Fourier series. The effects of the different terms are compounded with one another, provided the differential system linking the disturbances of pressure, density, velocity, etc., is linear. This condition is fulfilled here. If f_0 is the value of a parameter f in the reference state, we set $f = f_0 + f'$, and obtain

$$\begin{cases} p = p_0 + p', \rho = \rho_0 + \rho', T = T_0 + T', \\ \vec{\mathbf{v}} = \vec{\mathbf{v}}', \underline{\xi} = \underline{\xi_0} + \underline{\xi}', \underline{\mathbf{A}} = \underline{\mathbf{A}'}, s = s_0 + s' \end{cases} \quad [1.49]$$

We accept that the disturbances $p', \rho', T', \underline{\xi}', \vec{\mathbf{v}}'$ and their partial derivatives in relation to time and spatial coordinates are of the same order of magnitude, up to the order of three. This is the hypothesis of linearity. In addition, we accept that the propagation of sound is adiabatic.

As the medium is initially at chemical equilibrium, sound propagation causes a certain non-equilibrium. However, our theory remains valid for phenomena other than chemical relaxation, particularly when we see non-equilibria due to the internal DOFs of the molecules. We can indeed show that the exchanges between the translational DOFs and the internal DOFs of the molecules lead to the definition of relaxation times and can be dealt with in the same manner as in near-equilibrium chemical kinetics [KNE 65].

In addition, when different phenomena of transfer take place, we can show [MEI 59] that the absorption effect which occurs during sound propagation can be approximated as the sum of the absorption effects due to the contributions of the various transfer phenomena and internal DOFs such that we would obtain them were we to consider these phenomena separately.

This remark provides a justification for dealing with chemical relaxation alone.

Note finally that we can approach the propagation of small disturbances without making the hypothesis of linearity. Nonlinear sound propagation is a rather complex problem that can be solved by using asymptotic expansions for the different variables and their derivatives.

1.1.3.2. *Compressibility coefficient and dispersion relation*

For multi-reactive media, it may be advantageous to use the compressibility coefficients. Quite apart from the thermodynamic aspect, this is a useful step in solving the problem.[1]

If ρ' and p' represent the disturbances in the density and pressure, we use the symbol χ to denote the volumetric compressibility coefficient such that:

$$\chi = \lim_{\substack{\rho' \to 0 \\ p' \to 0}} \left(\rho' / \rho_0 p' \right) \qquad [1.50]$$

The determination of χ requires knowledge of the evolutions of ρ' and p'. Newton supposed that sound propagation was isothermal – a hypothesis which leads us to define, for a simple gas:

$$\chi_T = \frac{1}{\rho} \left(\frac{\partial \rho}{\partial p} \right)_T \qquad [1.51]$$

Through his experiments on air, Laplace showed that sound propagates adiabatically, so we are led to define the isentropic compressibility coefficient:

$$\chi_s = \frac{1}{\rho} \left(\frac{\partial \rho}{\partial p} \right)_s \qquad [1.52]$$

Definition of the coefficient χ_s also requires that it be possible to determine a relation between p and ρ for the medium and with the hypothesis of adiabaticity. This is possible in a chemically frozen medium, where we obtain:

$$\chi_s^f = \frac{1}{\rho} \left(\frac{\partial \rho}{\partial p} \right)_{s\underline{\xi}} \qquad [1.53]$$

which gives us the following for a fixed-composition mixture of perfect gases:

$$\chi_s^f = 1 / p \gamma_f \qquad [1.54]$$

1 We could have used them for a mono-reactive medium, but there is less of an advantage.

where γ_f is the isentropic coefficient:

$$\gamma_f = \left(\frac{\partial \ln p}{\partial \ln \rho}\right)_{s\underline{\xi}} = \frac{c_{pf}}{c_{vf}} = \frac{\sum_1^N n_j C_{pj}}{\sum_1^N n_j C_{pj} - Rn} \quad [1.55]$$

In a medium at chemical equilibrium, for which small disturbances do not cause disequilibrium, we obtain

$$\chi_s^e = \frac{1}{\rho}\left(\frac{\partial \rho}{\partial p}\right)_{s\underline{\mathbf{A}}} = \frac{\rho}{e_{\vartheta\vartheta} - \tilde{e}_{\vartheta\underline{\xi}}\, e_{\underline{\underline{\xi\xi}}}^{-1}\, e_{\xi\vartheta}} \quad [1.56]$$

and if we are dealing with a mixture of perfect gases:

$$\chi_s^e = 1/p\gamma_e \quad [1.57]$$

γ_e being the isentropic equilibrium coefficient, equal to

$$\gamma_e = \left(\frac{\partial \ln p}{\partial \ln \rho}\right)_{s\underline{\mathbf{A}}} = \frac{c_{pe}}{c_{pe}\left(1 - D_p\right) - Rn\left(1 + D_T\right)^2} \quad [1.58]$$

where

$$c_{pe} = \left(\partial h/\partial T\right)_{p,\underline{\mathbf{A}}},\ D_T = \left(\partial \ln n/\partial \ln T\right)_{p,\underline{\mathbf{A}}},\ D_p = \left(\partial \ln n/\partial \ln p\right)_{T,\underline{\mathbf{A}}} \quad [1.59]$$

When there is chemical relaxation, the definition of the compressibility coefficient is complicated by the non-existence of a thermodynamic relation between the density and pressure. We therefore define an effective compressibility coefficient χ_s^{eff} – a complex number whose real part is equal to the compressibility coefficient.

The continuity equation and momentum equation can be used to deduce a relation linking the values characteristic of the soundwave to the compressibility coefficient. Given equations [2.90] and [2.105] in [PRU 12], in the absence of diffusion, viscosity and remotely exercised forces, we have

$$\begin{cases} d\rho/dt + \rho\vec{\nabla}\cdot\vec{\mathbf{v}} = 0 \\ \rho\, d\,\vec{\mathbf{v}}/dt + \vec{\nabla}p = \vec{0} \end{cases} \qquad [1.60]$$

We linearize this using the equations from system [1.49], and in view of the fact that $\vec{\mathbf{v}}_0$ is zero, we obtain

$$\begin{cases} \partial\rho'/\partial t + \rho_0\,\vec{\nabla}\cdot\vec{\mathbf{v}}' = 0 \\ \rho_0\,\partial\,\vec{\mathbf{v}}'/\partial t + \vec{\nabla}p' = \vec{0} \end{cases} \qquad [1.61]$$

We consider the periodic solutions ρ', p' and $\vec{\mathbf{v}}'$.

As system [1.61] is linear, we can think in terms of complex variables and simply set

$$f' = \hat{f}\, e^{i\left(\omega t - \vec{\mathbf{K}}\cdot\vec{\mathbf{r}}\right)} \qquad [1.62]$$

where f' is the disturbance of the function f, ω the pulsation of the wave, $\vec{\mathbf{K}}$ the complex wavenumber vector, $\vec{\mathbf{r}}$ the position vector whose coordinates are the spatial variables and $\hat{f}$ is the maximum amplitude of the disturbance.

Equation [1.62] can be used to define the derivation operators for solutions of this type:

$$\partial/\partial t = i\,\omega, \vec{\nabla} = i\,\vec{\mathbf{K}} \qquad [1.63]$$

By eliminating the velocity $\vec{\mathbf{v}}'$ among the equations in system [1.61], we obtain

$$\vec{\mathbf{K}}\cdot\vec{\mathbf{K}}/\omega^2 = \rho'/p' \qquad [1.64]$$

This gives us the following formula for the complex compressibility coefficient:

$$\chi^{eff} = \frac{1}{\rho}\frac{K^2}{\omega^2} \qquad [1.65]$$

Equation [1.65] will yield the dispersion relation.

For a chemically frozen mixture, there are solutions where ω and K are real. The same is true for a mixture at chemical equilibrium for which the disturbances do not cause a relaxation phenomenon.

The ratio ω/K is therefore the velocity of the wave phase, i.e. the celerity of sound. From equations [1.53], [1.56] and [1.65], we obtain

$$\begin{cases} c_f^{\ 2} = \omega^2/K^2 = \vartheta^2\, e_{\vartheta\vartheta}, \text{(frozen mixture)} \\ c_e^{\ 2} = \omega^2/K^2 = \vartheta^2\left(e_{\vartheta\vartheta} - \tilde{e}_{\vartheta\underline{\xi}}\, e_{\underline{\xi\xi}}^{\ -1} e_{\vartheta\underline{\xi}}\right), \text{(mixture at} \\ \text{chemical equilibrium)} \end{cases} \qquad [1.66]$$

This lends acoustical meaning to the thermodynamic celerities of sound (characteristic velocities) c_e and c_f.

1.1.3.3. *Effective compressibility coefficient: dispersion relation in a multi-reactive medium*

The compressibility coefficient of a simple fluid in barotropic evolution is, by definition, $\chi = d\ln\rho/dp = -\rho\, d\vartheta/dp$. It is no longer possible to define such a thermodynamic coefficient in the case of a reactive fluid. Conversely, as we have seen, we can define an effective compressibility coefficient when exposed to small disturbances:

$$\chi^{eff} = -\rho_0\, \vartheta'/p' \qquad [1.67]$$

At near-equilibrium, the equation of species' production can be written as follows (instead of equation [1.28]):

$$\partial\underline{\xi}'/\partial t = \underline{\underline{\mathbf{L}}}\,\underline{\mathbf{A}} \qquad [1.68]$$

where $\underline{\underline{\mathbf{L}}}$ is a square, symmetrical, positive definite matrix (phenomenological coefficient matrix), $\underline{\xi}'$ is the vector whose components are the disturbances relating to the reduced progress variables.

The energy equation [2.122], presented in [PRU 12], can be replaced, using the Gibbs relation and the adopted hypotheses, with:

$$\frac{ds}{dt} - \frac{\underline{\underline{\tilde{\mathbf{A}}}}}{T}\frac{d\underline{\xi}}{dt} = 0 \qquad [1.69]$$

which gives us the following for small disturbances:

$$\frac{\partial s'}{\partial t} - \frac{\underline{\underline{\tilde{\mathbf{A}}_0}}}{T_0}\frac{d\underline{\xi}'}{dt} = 0 \qquad [1.70]$$

and because $\underline{\mathbf{A}_0}$ is zero, we can demonstrate the isentropy of the phenomenon, which is expressed as $\partial s'/\partial t = 0$. As the disturbance in entropy is zero at the initial instant and the reference state is characterized by an entropy independent of the spatial coordinates, we obtain

$$s' = 0 \qquad [1.71]$$

If we accept the hypothesis of linearity, the wave propagation is isentropic.

Given the thermodynamics of the medium, the disturbances in pressure and chemical affinity can be written as a function of the disturbances in the specific volume and progress variable:

$$\begin{cases} -p' = e_{\vartheta\vartheta}\vartheta' + \tilde{e}_{\vartheta\underline{\xi}}\underline{\xi}' \\ -\underline{\mathbf{A}} = e_{\underline{\xi}\vartheta}\vartheta' + e_{\underline{\underline{\xi\xi}}}\underline{\xi}' \end{cases} \qquad [1.72]$$

(instead of systems [1.32] and [1.33] in a mono-reactive medium).

In such conditions, equations [1.67], [1.68] and [1.71] and system [1.72] lead to a system of the following form, using the derivation operators in equation [1.63]:

$$\begin{cases} \vartheta' - \vartheta_0\left(e_{\vartheta\vartheta}\vartheta' + \tilde{e}_{\vartheta\underline{\xi}}\underline{\xi}'\right)\chi = 0, \\ z\underline{\xi}' + \underline{\underline{\mathbf{L}}}\left(e_{\underline{\xi}\vartheta}\vartheta' + e_{\underline{\underline{\xi\xi}}}\underline{\xi}'\right) = \underline{\mathbf{0}} \end{cases} \qquad [1.73]$$

This is a linear and homogeneous system of R+1 equations and R+1 unknowns: $\vartheta', \xi_1', \cdots \xi_R'$.

This system has no solution unless the associated determinant is zero. By eliminating ϑ' from system [1.73], we can deduce

$$\underline{\underline{\mathbf{L}}}\left[\left(1 - \vartheta_0 e_{\vartheta\vartheta}\chi\right)\left(z\underline{\underline{\mathbf{L}}}^{-1} + e_{\underline{\underline{\xi\xi}}}\right) + \vartheta_0\chi e_{\vartheta\underline{\xi}}\tilde{e}_{\underline{\xi}\vartheta}\right]\underline{\xi}' = \underline{\mathbf{0}} \qquad [1.74]$$

By particularizing z and χ, we obtain the following cases:

$$\begin{cases} a)\,\chi=\chi^{eff}\ ,\ z=i\,\omega, & b)\,\chi=\chi_s^e, z=0 \\ c)\,\chi=0\ ,\ z=-1/\tau_{s,\vartheta}, & d)\,\chi=\infty\ ,\ z=-1/\tau_{s,p} \end{cases} \qquad [1.75]$$

which correspond respectively to chemical relaxation, the hypothesis of chemical equilibrium, the chemical times at constant entropy and volume which are the inverses of the eigenvalues of the product of matrices $\underline{\underline{\mathbf{L}}}\,\underline{\underline{\mathbf{e}}}_{\xi\xi}$ (we shall accept that these chemical times are distinct), and to the chemical times at constant entropy and pressure.

As the matrix $\underline{\underline{\mathbf{L}}}$ is always regular, we can write the following, cancelling out the determinant from equation [1.74]:

$$\det\left[\left(1-\vartheta_0\, e_{\vartheta\vartheta}\,\chi\right)\left(z\,\underline{\underline{\mathbf{L}}}^{-1}+\underline{\underline{e}}_{\xi\xi}\right)+\vartheta_0\chi\, \underline{e}_{\vartheta\xi}\,\underline{\tilde{e}}_{\xi\vartheta}\right]=0 \qquad [1.76]$$

If we expand the left-hand side of it, equation [1.76] becomes

$$\begin{aligned}&\left(1-\vartheta_0\, e_{\vartheta\vartheta}\,\chi\right)^{\mathrm{R-1}}\Big[\left(1-\vartheta_0\, e_{\vartheta\vartheta}\,\chi\right)\det\left(z\,\underline{\underline{\mathbf{L}}}^{-1}+\underline{\underline{e}}_{\xi\xi}\right)+\\ &\vartheta_0\chi\, \underline{e}_{\vartheta\xi}\,\mathrm{Adj}\left(z\,\underline{\underline{\mathbf{L}}}^{-1}+\underline{\underline{e}}_{\xi\xi}\right)\underline{\tilde{e}}_{\xi\vartheta}\Big]=0\end{aligned} \qquad [1.77]$$

This equation gives us the solution

$$1-\vartheta_0\, e_{\vartheta\vartheta}\,\chi=0 \qquad [1.78]$$

i.e. $\chi=\chi_s^f$, which characterizes media with fixed composition, and the relaxation solution:

$$\chi=\frac{\vartheta_0^{-1}\det(z\,\underline{\underline{\mathbf{L}}}^{-1}+\underline{\underline{e}}_{\xi\xi})}{e_{\vartheta\vartheta}\det(z\,\underline{\underline{\mathbf{L}}}^{-1}+\underline{\underline{e}}_{\xi\xi})-\underline{\tilde{e}}_{\vartheta\xi}\,\mathrm{Adj}(z\,\underline{\underline{\mathbf{L}}}^{-1}+\underline{\underline{e}}_{\xi\xi})\,\underline{e}_{\xi\vartheta}} \qquad [1.79]$$

The numerator and denominator of the right-hand side of equation [1.79] are polynomials of degree R in z such that χ assumes the form

$$\chi = \chi_0 \prod_1^R (1 + a_r z) \Big/ \prod_1^R (1 + b_r z) \qquad [1.80]$$

In view of equation [1.75b], we obtain χ_s^e by making $z = 0$ so that $\chi_0 = \chi_s^e$.

Given equation [1.75c], $\chi = 0$ leads to the R values $z = -1/\left(\tau_{s,\vartheta}\right)_r$ and according to equation [1.75d], $\chi = \infty$ gives the R values $z = -1/\left(\tau_{s,p}\right)_r$ which are respectively the roots of the polynomials of the numerator and denominator of the right-hand side of equation [1.80], so that we find $a_r = \left(\tau_{s,\vartheta}\right)_r$ and $b_r = \left(\tau_{s,p}\right)_r$.

Finally, by replacing χ with χ_s^{eff} (the subscript s denoting that it is an isentropic compressibility coefficient, as we have demonstrated) and z with $i\omega$, we obtain the expression of the complex effective compressibility coefficient.

This result means that we can write the dispersion relation in the form

$$\frac{K^2}{\omega^2} = \frac{1}{c_e^{\,2}} \frac{\prod_1^R \left(1 + i\omega\left(\tau_{s,\vartheta}\right)_r\right)}{\prod_1^R \left(1 + i\omega\left(\tau_{s,p}\right)_r\right)} \qquad [1.81]$$

because $\rho_0 \chi_s^{eff} = K^2/\omega^2$ and $\lim_{\omega \to 0}\left(K^2/\omega^2\right) = \rho_0 \chi_s^e = 1/c_e^{\,2}$.

This equation can be solved by separating the real and imaginary parts of both sides.

The column matrix representing the wavenumber vector $\underline{\mathbf{K}}$ is written:

$$\underline{\mathbf{K}} = \underline{\mathbf{k}} - i\,\underline{\boldsymbol{\gamma}} \qquad [1.82]$$

with $\underline{\mathbf{k}}$ and $\underline{\boldsymbol{\gamma}}$ being two real co-linear vectors in the physical space, so the modulus of $\underline{\mathbf{K}}$ is:

$$K = k - i\gamma = \omega/c - i\gamma \qquad [1.83]$$

with c being the celerity of sound and γ the absorption coefficient.

The form [1.81] of the dispersion relation is very easy to use because we need only know $c_e^2 = \rho_0 \chi_s^e$, which is a thermodynamic parameter, and the chemical times with constant s and ϑ and constant s and p, which simply necessitates the calculation of eigenvalues whose sign is known.

Note also with equation [1.81] that if $\omega \to \infty$, the celerity of sound tends toward that of the medium with constant concentrations c_f. Therefore, we have

$$\frac{c_e^2}{c_f^2} = \prod_1^R \frac{\left(\tau_{s,\vartheta}\right)_r}{\left(\tau_{s,p}\right)_r} \qquad [1.84]$$

1.1.3.4. *Second form of the dispersion relation*

We can express the effective compressibility coefficient in a simple canonical form by revealing relaxation forces.

To do this, we need only decompose into simple elements the rational fraction on the right-hand side of [1.81]. We replace $i\omega$ with z, giving us ([PRU 69]):

$$\frac{\prod_1^R \left(1 + z\left(\tau_{s,\vartheta}\right)_r\right)}{\prod_1^R \left(1 + \left(\tau_{s,p}\right)_r\right)} = \frac{c_e^2}{c_f^2} + \sum_1^R \frac{\varepsilon_r}{1 + z\left(\tau_{s,p}\right)_r} \qquad [1.85]$$

By making $z = 0$ in equation [1.85], we get

$$\sum_1^R \varepsilon_r = 1 - \frac{c_e^2}{c_f^2} = \varepsilon > 0 \qquad [1.86]$$

in accordance with the stability conditions of thermodynamics.

We can show that the coefficients ε_r are all positive [PRU 69], and we shall call them relaxation forces associated with each chemical DOF of the medium.

The complex effective compressibility coefficient becomes

$$\chi_s^{eff} = \chi_s^e \left[1 - \sum_1^R \varepsilon_r \frac{i\omega\left(\tau_{s,p}\right)_r}{1 + i\omega\left(\tau_{s,p}\right)_r} \right] \qquad [1.87]$$

After separation of the real and imaginary parts, the dispersion relation becomes

$$\begin{cases} \dfrac{1}{c^2} - \dfrac{\gamma^2}{\omega^2} = \dfrac{1}{c_e^2} \left[1 - \displaystyle\sum_1^R \varepsilon_r \frac{\omega^2 \left(\tau_{s,p}\right)_r^2}{1 + \omega^2 \left(\tau_{s,p}\right)_r^2} \right] \\ \dfrac{2\gamma}{\omega c} = \dfrac{1}{c_e^2} \displaystyle\sum_1^R \varepsilon_r \frac{\omega \left(\tau_{s,p}\right)_r}{1 + \omega^2 \left(\tau_{s,p}\right)_r^2} \end{cases} \qquad [1.88]$$

1.1.3.5. *Special cases*

When the force of relaxation ε is far smaller than one unit, we need only set $\varepsilon_r = \varepsilon\,\delta_r$ and expand this in terms of ε. We obtain

$$\begin{cases} c^2 \cong c_e^2 + \displaystyle\sum_{r=1}^R \frac{\left(c_f^2 - c_r^2\right) \omega^2 \left(\tau_{s,p}\right)_r^2}{1 + \omega^2 (\tau_{s,p})_r^2} \\ \gamma \cong \displaystyle\sum_{r=1}^R \frac{\left(c_f^2 / c_r^2 - 1\right)}{2 c_f (\tau_{s,p})_r} \frac{\omega^2 \left(\tau_{s,p}\right)_r^2}{1 + \omega^2 (\tau_{s,p})_r^2} \end{cases} \qquad [1.89]$$

defining the thermodynamic celerity c_r by the relation

$$\varepsilon_r = 1 - c_r^2 / c_f^2 \qquad [1.90]$$

In this case, only the values c^2 and γ are equal to the sums of the contributions of each DOF, as if the DOFs were independent.

The principle of superposition applies.

Figure 1.2 shows the dispersion and absorption curves for a mixture with two chemical DOFs where $\left(\tau_{s,p}\right)_2 << \left(\tau_{s,p}\right)_1$.

In this case, we have

$$c^2 \cong c_e^{\,2} + \left(c_f^{\,2} - c_1^{\,2}\right)\frac{\omega^2\left(\tau_{s,p}\right)_1^2}{1+\omega^2\left(\tau_{s,p}\right)_1^2} + \left(c_f^{\,2} - c_2^{\,2}\right)\frac{\omega^2\left(\tau_{s,p}\right)_2^2}{1+\omega^2\left(\tau_{s,p}\right)_2^2} \qquad [1.91]$$

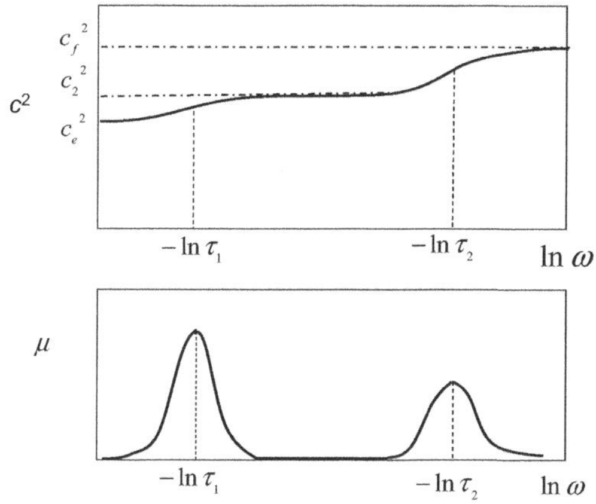

Figure 1.2. *Dispersion and absorption curves when τ_1 and τ_2 are of different orders of magnitude*

Using the coefficient of absorption per unit wavelength $\mu = 2\pi\gamma/\omega c$ we obtain

$$\mu \cong \frac{\pi}{a_e^{\,2}}\left[\frac{\varepsilon_1\,\omega\left(\tau_{s,p}\right)_1}{1+\omega^2\left(\tau_{s,p}\right)_1^2} + \frac{\varepsilon_2\,\omega\left(\tau_{s,p}\right)_2}{1+\omega^2\left(\tau_{s,p}\right)_2^2}\right] \qquad [1.92]$$

The evolution of μ is represented in Figure 1.3 in the case of a mixture of hydrogen and fluorine for which $\left(\tau_{s,p}\right)_1$ and $\left(\tau_{s,p}\right)_2$ are of the same order of magnitude.

When the relaxation force ε is no longer very small in comparison to one unit, we can no longer apply the principle of superposition. However, we have the following results:

– the celerity of sound c is an increasing function of the wave pulsation ω. When this is very small in relation to the inverses of the chemical times, c is equal to c_e. When ω is very large, c is equal to c_f. The dispersion curve may exhibit staggered platforms, as shown in Figure 1.2, if the chemical times are of different orders of magnitude;

– the absorption curve $\mu(\omega)$ exhibits one or more maxima. The value μ tends toward zero when the frequency is zero or infinite. The absorption coefficient γ tends toward zero for a zero frequency and toward a positive limit for an infinite frequency.

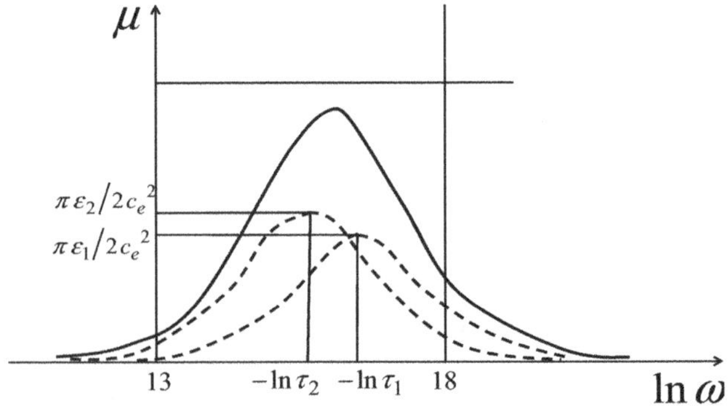

Figure 1.3. *Absorption curve when τ_1 and τ_2 are of the same order of magnitude (hydrogen and fluorine mixture) with $1-c_e^2/c_f^2 \ll 1$. Solid line: absorption curve. Dotted lines: curves corresponding to each of the two terms in equation [1.92]*

1.1.3.6. *Wave equation*

Equation [1.81] was obtained by using the derivation operator $i\omega$ for $\partial/\partial t$. Had we taken the direct route, we would have obtained the following by successive derivations:

$$c_e^2 \prod_1^R \left[1+\left(\tau_{s,p}\right)_r \partial/\partial t\right]\rho' = \prod_1^R \left[1+\left(\tau_{s,\vartheta}\right)_r \partial/\partial t\right]p' \qquad [1.93]$$

The momentum equation enables us to set

$$\vec{\mathbf{v}}' = \vec{\nabla}\varphi' \text{ and } p' = -\rho_0\, \partial\varphi'/\partial t$$

so the continuity equation becomes

$$\partial\rho'/\partial t + \rho_0 \nabla^2 \varphi' = 0 \qquad [1.94]$$

Eliminating ρ' and p' between relations [1.93] and [1.94] gives us the wave equation:[2]

$$\frac{1}{c_e^2}\prod_1^R\left[1+\left(\tau_{s,\vartheta}\right)_r\frac{\partial}{\partial t}\right]\frac{\partial^2\varphi'}{\partial t^2}-\prod_1^R\left[1+\left(\tau_{s,p}\right)_r\frac{\partial}{\partial t}\right]\nabla^2\varphi'=0 \qquad [1.95]$$

A second form of the wave equation is obtained by using equation [1.87], so that by introducing the relaxation forces ε_r or the thermodynamic celerities of sound c_r (also known as characteristic celerities), we obtain

$$\prod_1^R\left[1+\left(\tau_{s,p}\right)_r\frac{\partial}{\partial t}\right]\nabla^2\varphi'=$$

$$\frac{1}{c_e^2}\left\{\prod_1^R\left[1+\left(\tau_{s,p}\right)_r\frac{\partial}{\partial t}\right]-\sum_{i=1}^R\left(1-\frac{c_r^2}{c_f^2}\right)\left(\tau_{s,p}\right)_r\frac{\partial}{\partial t}\prod_{j\neq r}\left[1+\left(\tau_{s,p}\right)_j\frac{\partial}{\partial t}\right]\right\}\frac{\partial^2\varphi'}{\partial t^2}$$

[1.96]

1.1.3.7. *Mixture with a single chemical degree of freedom (DOF)*

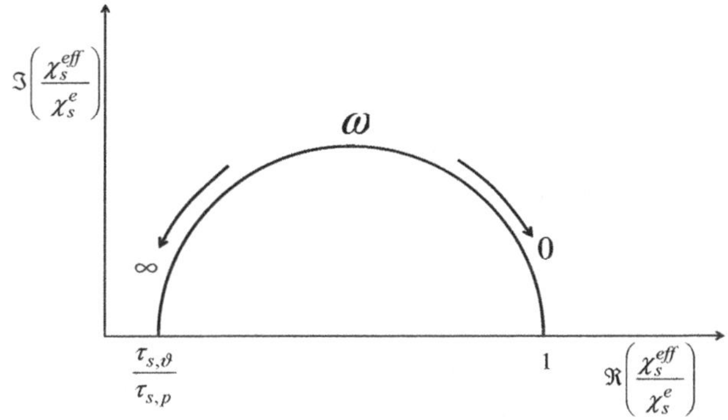

Figure 1.4. *Variations of the complex reduced effective compressibility coefficient with the wave pulsation* ω

---

2 NOTE.– This wave equation applies to plane, cylindrical or spherical waves. We need only set $\nabla^2=\frac{1}{r^{n-1}}\frac{\partial}{\partial r}\left(r^{n-1}\frac{\partial}{\partial r}\right)$, with the three cases:

$n=1$, plane wave, $r=x$;
$n=2$, cylindrical wave, r being the distance from the axis;
$n=3$ spherical wave, r being the distance from the center.

Using the above analysis and setting R as equal to 1, we of course find the same results as with mono-reactive mixtures found in section 1.1.2.

In particular, we obtain the following formulae for the complex effective compressibility coefficient χ_s^{eff} introduced here in section 1.2 and for the relative difference ε of the equilibrium- and frozen celerities of sound:

$$\chi_s^{eff} = \chi_s^e \frac{1+i\omega\tau_{s,\vartheta}}{1+i\omega\tau_{s,p}}, \ \varepsilon = 1-\frac{\tau_{s,\vartheta}}{\tau_{s,p}} = 1-\frac{c_e 2}{c_f^{\,2}} \quad [1.97]$$

When ε is much smaller than one unit, the celerity of sound and the absorption coefficient are given by system [1.47], and the wave equation is given by equation [1.40].

Figure 1.4 shows the evolutions of χ_s^{eff} / χ_s^e, when ω varies from 0 to $+\infty$.

The curve formed is a semi-circle with diameter ε.

1.2. Relaxed flows in nozzles

Quite apart from its own inherent interest, section 1.1, dedicated to sound propagation, can be applied directly to the nozzles of thrusters. Indeed, it deals with the question of de Laval nozzles, and we saw in section 1.6.6 of [PRU 12] that the regimes of operation of these nozzles were governed by the Mach number of the flow, which is the ratio of the speed of that flow to the speed of sound. However, in the case studied in Chapter 1, there were no chemical reactions.

This is no longer true with what generally happens in thruster nozzles located at the exit from the engine. The high temperatures encountered at the opening of nozzles and the cooling experienced during the expansion of the gases mean that the combustion products undergo chemical reactions of recombination. As we saw in section 1.1, the speed of sound, at a given temperature and pressure, is no longer unique in a reactive medium and therefore can no longer serve as a reference point to define the conditions at the throat in the adapted regime. A 1D study using asymptotic expansions gives us a clearer picture of what happens at the throat.

Finally, we shall extend the study of flows in nozzles to the case of a fluid mixture exhibiting non-equilibria of the internal modes of energy in addition to the chemical reactions. This opens the door for a discussion of the re-entry of bodies into the atmosphere at high speeds, although we shall overlook the ionization that occurs.

1.2.1. *Calculation of a continuous flow with a recombination-dissociation reaction in a de Laval nozzle*

Suppose that the dissipative fluxes $\vec{\jmath}_{Dj}$, $\vec{\vec{\mathbf{P}}} - p\vec{\vec{\mathbf{1}}}$ and $\vec{\mathbf{q}}$ are null, as are r and $\vec{f}_j$. The rate of entropy production is therefore equal to

$$\dot{W}_S = -\sum_j \frac{g_j}{T}\dot{W}_j \qquad [1.98]$$

As pointed out in Chapter 2 of [PRU 12], a knowledge of the laws of chemical kinetics means that we do not have to use linearized laws to express the rate of chemical production. The hypothesis of chemical near-equilibrium is therefore not necessary, and the balance equations give us a way to solve problems of flow in circumstances very far removed from chemical equilibrium [GUR 65, NAP 63].

The balance equations become

$$\rho\, d\vartheta/dt = \vec{\nabla}\cdot\vec{\mathbf{v}} \qquad [1.99]$$

$$\rho\, dY_j/\partial t = \dot{W}_j \qquad [1.100]$$

$$\rho\, d\,\vec{\mathbf{v}}/dt + \vec{\nabla}p = 0 \qquad [1.101]$$

$$\rho\, de/dt = -p\vec{\nabla}\cdot\vec{\mathbf{v}} \qquad [1.102]$$

Elimination of $\vec{\nabla}\cdot\vec{\mathbf{v}}$ gives us

$$de/dt + p\, d\vartheta/dt = 0 \qquad [1.103]$$

or, according to the Gibbs relation

$$\rho\, ds/dt + \sum_j g_j dY_j/dt = 0 \qquad [1.104]$$

In a steady flow, we have

$$d\left(h + \mathbf{v}^2/2\right)/dt = 0 \qquad [1.105]$$

The total energy balance equation becomes

$$h+\mathbf{v}^2/2=h_0 \quad [1.106]$$

If, in addition, the chemical process is reduced to a single reversible reaction, such as

$$A_2+M \rightleftarrows A+A+M \quad [1.107]$$

where M is a neutral diluent or one of the species A or A_2, we have

$$\begin{cases} \dot{W}_A = 2\mathcal{M}_A\left(\dot{\zeta}_D-\dot{\zeta}_R\right) \\ \dot{W}_{A_2} = -\mathcal{M}_{A_2}\left(\dot{\zeta}_D-\dot{\zeta}_R\right) \end{cases} \quad [1.108]$$

where

$$\begin{cases} \dot{\zeta}_D = k_D(T)C_{A_2}C,\ \dot{\zeta}_R = k_R(T)C_A{}^2C \\ \mathcal{M}_{A_2} = 2\mathcal{M}_A,\ C=\sum_j C_j \end{cases} \quad [1.109]$$

We set

$$dY_j = \nu_j\mathcal{M}_j\,d\xi \quad [1.110]$$

From this, it follows that

$$\rho\,d\xi/dt = \dot{\zeta}_D-\dot{\zeta}_R \quad [1.111]$$

The system needing to be solved therefore becomes

$$\begin{cases} \vec{\nabla}\cdot\vec{\mathbf{v}}=0 \\ \rho\,ds/dt - A\,d\xi/dt = 0 \\ \rho\,d\xi/dt = \dot{\zeta}_D-\dot{\zeta}_R \\ h+\vec{\mathbf{v}}^2/2=h_0 \\ A=-\sum_j \nu_j\mathcal{M}_j g_j \end{cases} \quad [1.112]$$

Obviously, we need to take account of the equations of state:

$$p = p(T,\rho,\xi),\ h = h(T,\rho,\xi),\ s = s(T,\rho,\xi) \qquad [1.113]$$

The simplest case is that of *flow in planar sections* (mono-dimensional or quasi-one-dimensional flow). We accept that the parameters are uniform in every cross-section of the streamtube studied, as done in section 1.6.6 of [PRU 12].

Then, a single variable remains: the abscissa x along the middle stream line. If $S(x)$ is the surface area of the normal cross-section and if U is the average velocity in that cross-section, we obtain the following system:

$$\begin{cases} \rho U\, S(x) = \dot{m},\ \rho\, ds/dx - A\, d\xi/dx = 0 \\ \rho U\, d\xi/dt = \dot{\zeta}_D - \dot{\zeta}_R,\ h + U^2/2 = h_0 \\ p = p(T,\rho,\xi),\ h = h(T,\rho,\xi),\ s = s(T,\rho,\xi) \end{cases} \qquad [1.114]$$

This system can be solved numerically, and the solution demonstrates the coupling between the flow and the chemistry. In a rocket thruster nozzle[3], the flow is generally subsonic and then supersonic (Figure 1.5).

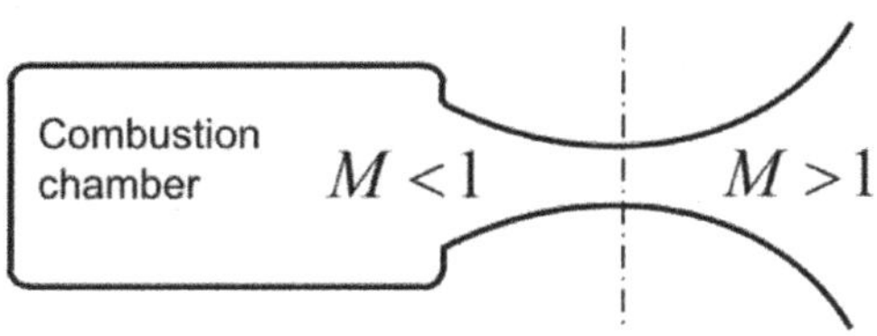

Figure 1.5. *Architecture of a rocket nozzle*

The temperature, pressure and concentration Y_A are decreasing functions of the abscissa x, the concentration of A_2 increases because of the recombination when the temperature drops. The rates of chemical production $\dot{\zeta}_D$ and $\dot{\zeta}_R$ have the shapes illustrated in Figure 1.6:

3 Mono-dimensional flow in a de Laval nozzle is discussed in Chapter 1 of [PRU 12] for a non-reactive ideal gas. Here, the evolution is complicated by the chemical reactions that occur; however, we shall preserve the same notation wherever possible.

– for $x < a$, we have $\dot{\zeta}_D / \dot{\zeta}_R \cong 1$, chemical equilibrium is practically reached and $A \cong 0$. The flow is relatively slow and at high temperature, the gradients are slight: $ds/dx \cong 0$;

– for $a < x < b$, the evolution is a very long way from equilibrium, and the temperature is still sufficiently high for the reactions to take place, but they are less quick: $ds/dx > 0$;

– for $x > b$, the reaction rates are slow, but the tendency for recombination remains greater than the tendency for dissociation of the species A_2. We have $\dot{\zeta}_D / \dot{\zeta}_R \cong 0$ and a very small $\rho U \, d\xi/dx \cong -\dot{\zeta}_R$. The flow is frozen and we have $ds/dx \cong 0$.

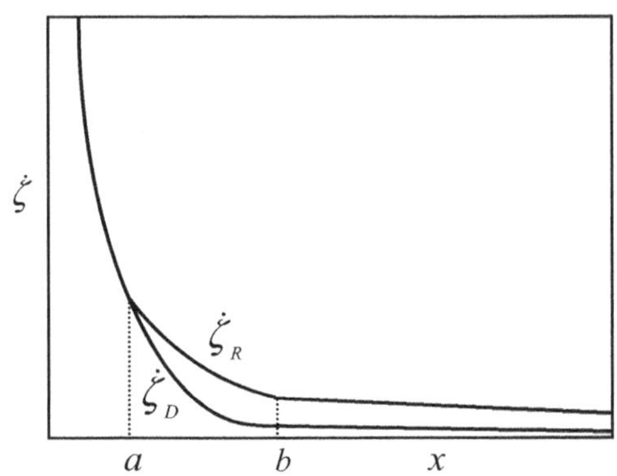

Figure 1.6. *Rates of chemical production in a rocket nozzle*

In any case, the rate of entropy production is equal to

$$\dot{W}_S = \dot{\zeta}_R \, A/T \left(e^{A/RT} - 1\right) \qquad [1.115]$$

In certain cases, when the irreversible zone (a b) is very small, we can assimilate the evolution to the succession of a flow at chemical equilibrium and a fixed flow. There is sudden freezing at point a where the rate $\dot{W}_S$ can be assimilated to a Dirac peak: $\dot{W}_S = \dot{W}_{Sa} \, \delta(x - a)$.

The chemical irreversibilities (chemical relaxation) limit the recombination of species A in the flow; although the reaction is exothermic, the energy given off by it is less than that given off in the equilibrium state. The thrust of the rocket engine is therefore diminished by the phenomenon of relaxation. In order to evaluate the extent of chemical irreversibility in advance, we use the first Damköhler parameter $Da = \tau_{\text{mech}} / \tau_{\text{chem}}$ before solving the above system numerically.

Another difference with non-reactive flow is the behavior at the throat of the nozzle. Indeed, unlike what was seen in section 1.6.6 of [PRU 12], the velocity of the flow at the throat is no longer equal to the celerity of sound.

The difference in behavior becomes clearer still if we analyze what happens whilst taking account of the two-dimensional nature of the flow.

Therefore, let us turn out attention to the transonic zone using the asymptotic expansion method in the first hypothesis.[4]

1.2.2. *Asymptotic study of the transonic zone of a continuous mono-dimensional flow in a de Laval nozzle*

1.2.2.1. *Recap of the results relating to the transonic flow of a non-reactive ideal gas*

In section 1.6.6 of [PRU 12], we saw that we could calculate the parameters of the flow of a *non-reactive* ideal gas as a function of one of them, even in the transonic zone. Adjustment in a given de Laval nozzle takes place, in subsonic-supersonic evolution, by way of the conservation of the mass flowrate and the conditioning at the throat of the nozzle where the flow is sonic, in accordance with equation [1.73], given again here: $\frac{dp}{dx} = \frac{\rho U^2}{M^2 - 1} \frac{1}{S} \frac{dS}{dx}$. Let us use the small disturbance method in the vicinity of the basic flow defined by $\vartheta_0, U_0 = c_0, s_0, S_0, \dot{m}_0$.

We shall use the same notations with the dimensionless parameters adjusted to that basic flow. We choose a small parameter ε characterizing the variations of the normal section of the nozzle in the vicinity of the throat and consider a strictly positive number n to be determined. The expansions are as follows:

---

4 Two-dimensional study is more complex and will not be discussed here. It brings into play the transonic characteristics [NAP 66; PRU 69; PRU 70; PRU 10].

$S=1+\varepsilon S_1$, $\dot{m}=1+\varepsilon \dot{m}_1+\varepsilon^2 \dot{m}_2+\ldots$, $s=1+\varepsilon s_1+\varepsilon^2 s_2+\ldots$, $\vartheta=1+\varepsilon \vartheta_1+\varepsilon^2 \vartheta_2+\ldots$, $U=1+U_1+\varepsilon^2 U_2+\ldots$. We find that for $n=1/2$:

$$\frac{\gamma+1}{2}U_1^2-S_1+\dot{m}_2=0,\ \vartheta_1=U_1, s_1=0,\ \dot{m}_1=0 \qquad [1.116]$$

The profile of the nozzle, supposed to be parabolic, being given by $S_1=S_1{}^*+ax^2$, this leads to $\frac{\gamma+1}{2}U_1^2-ax^2=S_1{}^*-\dot{m}_2$. The solution of the problem in the plane (x,U_1) is a family of hyperbolas. The critical mass flowrate is that which renders the right-hand side $S_1{}^*-\dot{m}_2$ of the above equation equal to zero, which corresponds to the subsonic-supersonic or supersonic-subsonic continuous flow (Figure 1.7). If the mass flowrate is less than the critical value, we also have a continuous flow, but the regime is either subsonic or supersonic. At a local level we see the result shown in Figure 1.15 of [PRU 12].

1.2.2.2. *Transonic flow of a mono-reactive gas*

In a reactive flow, the singular point moves downstream of the throat. It is possible to analytically study the transonic zone using the singular small disturbance method, as we did previously for the case of a non-reactive fluid.

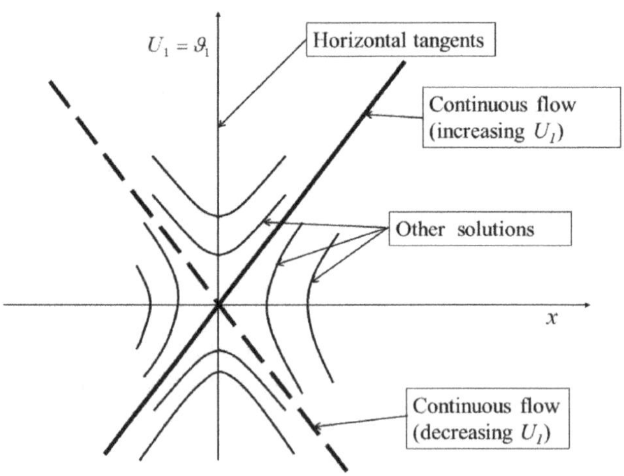

Figure 1.7. *Transonic behavior of a simple fluid in a de Laval nozzle*

In order to do this, we make the simplifying hypothesis that has already been used for sound propagation, whereby the characteristic celerities c_f and c_e are near to one other, so $c_f^{\,2}/c_e^{\,2}-1=e_{\vartheta\xi}^{\,2}/\left(e_{\vartheta\vartheta}e_{\xi\xi}-e_{\vartheta\xi}^{\,2}\right)<<1$. The reference flow, which is uniform and at chemical equilibrium, is characterized by the values $A_0=0$ and $\vartheta_0, U_0=c_{f0}, s_0, S_0, \dot{m}_0, \xi_0$, where $\xi_0=-(\partial\xi/\partial\vartheta)_{sA,0}$. We also introduce a chemical time $(\tau_{s\vartheta})_0$ and the ratio $\theta=\tau_r/\tau_c$ where τ_r is a reference time and the chemical length $l_c=U_0\tau_c$. Without changing the notations for the dimensionless values, system [1.114] becomes

$$US=\dot{m}\vartheta,\ T\partial s/\partial x-A\partial\xi/\partial x=0,\ U\partial\xi/\partial x=\theta L A,\ h+U^2/2=h_0$$

The phenomenological coefficient L results from the linearization of the rate of chemical production near to chemical equilibrium. In addition to the expansions $S=1+\varepsilon S_1$, $\dot{m}=1+\varepsilon\dot{m}_1+\varepsilon^2\dot{m}_2+\ldots$, $s=1+\varepsilon s_1+\varepsilon^2 s_2+\ldots$ and $\vartheta=1+\varepsilon\vartheta_1+\varepsilon^2\vartheta_2+\ldots$, $U=1+U_1+\varepsilon^2 U_2+\ldots$, we would have

$$\xi=\xi_0+\varepsilon^n\xi_1+\varepsilon^{2n}\xi_2+\ldots \text{ and } A=\varepsilon^n A_1+\varepsilon^{2n}A_2+\ldots.$$

The first order of chemical affinity can be deduced from the fundamental given energy law: $e=e(s,\vartheta,\xi)$: $A_1=-e_{\xi s0}s_1-e_{\xi\vartheta 0}\vartheta_1-e_{\xi\xi 0}\xi_1$.

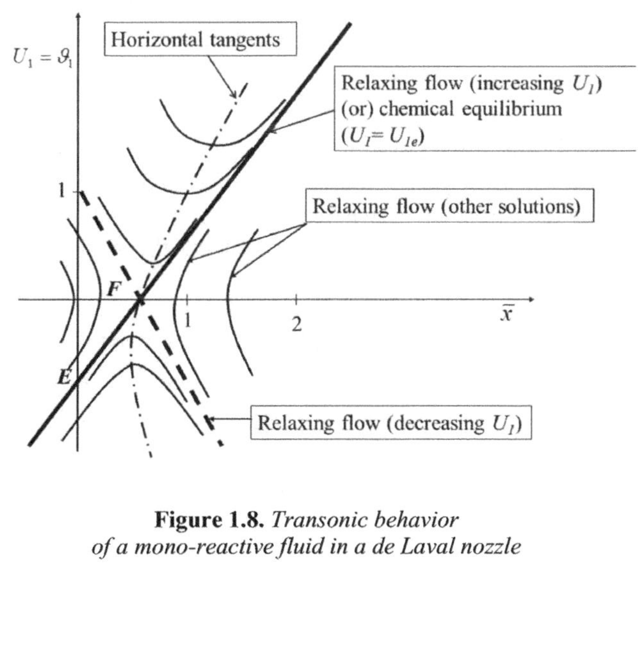

Figure 1.8. *Transonic behavior of a mono-reactive fluid in a de Laval nozzle*

At order 2, $A_2 = -e_{\xi s0} s_2 - e_{\xi \vartheta 0} \vartheta_2 - e_{\xi\xi 0} \xi_2 + \frac{A_{ss0}}{2} s_1^2 + \frac{A_{\vartheta\vartheta 0}}{2} \vartheta_1^2 + \frac{A_{\xi\xi 0}}{2} \xi_1^2$ $+ A_{s\vartheta 0} s_1 \vartheta_1 + A_{s\xi 0} s_1 \xi_1 + A_{\vartheta\xi 0} \vartheta_1 \xi_1$.

We write $\theta = \tau_r / \tau_c = e^{\delta n}$, and also introduce an arbitrary number α which facilitates a change of the spatial scale.

Thus, $\bar{x} = \varepsilon^{\alpha n} x = \varepsilon^{\alpha n} x^+ / l_r = \varepsilon^{(\delta - \alpha) n} x^+ / l_c$ the exponent + designator of the natural scale. Here, we shall not write the system that needs to be solved in general, because primarily what we want is the result obtained by applying the principle of least degeneracy, which leads us to choose $n = 1/2, \delta = \alpha$, so $\bar{x} = x^+ / l_c$. The final system to be solved is

$$Z \xi_1 = \frac{\Gamma + 1}{2} U_1^2 - S_1 + \dot{m}_2, \frac{d\xi_1}{d\bar{x}} + \xi_1 + \vartheta_1 = 0, \vartheta_1 = U_1, s_1 = 0, \dot{m}_1 = 0 \quad [1.117]$$

where $\Gamma = h_{\vartheta\vartheta 0}, c_{f0}^2 / c_{e0}^2 = Z\varepsilon^{1/2}, U_0 = c_{f0} = \vartheta_0 = 1, \tau_c = (\tau_{s\vartheta})_0$. We solve system [1.117] by first eliminating ξ_1. We obtain a differential equation in terms of U_1, the right-hand side of which depends on the profile of the throat and on the mass flowrate.

Figure 1.8 shows the solution obtained for $U_1 = \vartheta_1$ as a function of *x* in the case of a parabolic profile of the throat and for the following values of $\dot{m}_2$ and $\dot{m}_{2e}$: $\dot{m}_2 = S_1 * +1.25$; $\dot{m}_{2e} = S_1 * +0.25$.

We note significant differences with the case of a simple fluid shown in Figure 1.7: the existence of two particular points *E* and *F*, but the singularity is downstream from the throat at *F*, whereas it is at the point of the throat in the absence of chemical reaction. This difference renders the system more difficult to resolve [NAP 66].

The profile of the concentrations (progress variable) is not shown here. It can be consulted in [PRU 69; PRU 70; PRU 10].

1.3. Flows in thermal and chemical non-equilibrium

High-temperature reactive gaseous mixtures may experience not only chemical non-equilibria, but also non-equilibria of the internal DOFs of the molecules (translation, rotation, vibration). These non-equilibria manifest themselves in the boundary layers of re-entry bodies, for instance, but they are also to be seen in the flows of air in an arc nozzle for instance, where the fluid is raised to temperatures approaching 10,000 K. When they expand, the gases relax and cause the apparition of temperatures specific to the molecules and to each of their DOFs. However, we neglect the ionized species [DEL 63].

1.3.1. *Balance equations and closure relations in the presence of thermal and chemical non-equilibria*

1.3.1.1. *Laws of state*

Consider a composite mixture with multiple internal DOFs and with different translational temperatures [CLA 64; PRU 88; WOO 75; VIN 86], i.e. mutual non-equilibria [NAP 71].

The Gibbs relation is written thus for each sub-system comprising a species α:

$$\begin{cases} de_{\alpha,t} = T_{\alpha,t}\, ds_{\alpha,t} - p_\alpha d\vartheta_\alpha \\ de_{\alpha,\text{int}} = T_{\alpha,\text{int}}\, ds_{\alpha,\text{int}} \end{cases} \quad [1.118]$$

where the subscripts refer: t to translational mode, "int" to the internal modes and "$\text{int} \neq t$" to the internal modes except for translational motion. The quantities $\vartheta_\alpha = 1/\rho_\alpha$, p_α and T_α are, respectively, the specific volume, the partial pressure and the temperature of species α in the mixture. Assuming the gases to be perfect, we also have

$$p_\alpha = \rho_\alpha \frac{R}{\mathcal{M}_\alpha} T_{\alpha,t}, \quad e_\alpha = \sum_{\text{int}} e_{\alpha,\text{int}} + h_\alpha^0 \quad [1.119]$$

where h_α^0 represents the standard enthalpy of formation of species α, and e_α represents its mass internal energy.

Given that the internal energies and entropies are additive, it follows that (Y_α is the mass fraction of α):

$$e=\sum_{\alpha,\text{int}} Y_\alpha e_{\alpha,\text{int}}, \quad s=\sum_{\alpha,\text{int}} Y_\alpha s_{\alpha,\text{int}} \qquad [1.120]$$

According to statistical mechanics (see Appendix 2), applied in a generalized form to the present composite system, the thermodynamic functions are written as follows:

$$\begin{cases} e_{\alpha,t}\left(T_{\alpha,t}\right)=\dfrac{R}{\mathcal{M}_\alpha}T_{\alpha,t}{}^2\dfrac{\partial \ln\left[Q_{\alpha,t}\left(T_{\alpha,t}\right)\right]}{\partial T_{\alpha,t}}, \quad e_{\alpha,\text{int}\neq t}\left(T_{\alpha,\text{int}\neq t}\right) \\ =\dfrac{R}{\mathcal{M}_\alpha}T_{\alpha,\text{int}}{}^2\dfrac{\partial \ln\left[Q_{\alpha,\text{int}}\left(T_{\alpha,\text{int}}\right)\right]}{\partial T_{\alpha,\text{int}}} \\ s_{\alpha,t}\left(T_{\alpha,t}\right)=\dfrac{R}{\mathcal{M}_\alpha}\ln\left[Q_{\alpha,t}\left(T_{\alpha,t}\right)+1\right]+\dfrac{e_{\alpha,t}\left(T_{\alpha,t}\right)}{T_{\alpha,t}}, \\ s_{\alpha,\text{int}\neq t}\left(T_{\alpha,\text{int}\neq t}\right)=\dfrac{R}{\mathcal{M}_\alpha}\ln\left[Q_{\alpha,\text{int}}\left(T_{\alpha,\text{int}}\right)\right]+\dfrac{e_{\alpha,\text{int}}\left(T_{\alpha,\text{int}}\right)}{T_{\alpha,\text{int}}} \end{cases} \qquad [1.121]$$

1.3.1.2. *Balance laws*

We obtain equations of the same type as those in Chapter 2 of [PRU 12], taking account of the existence of the sub-systems:

$$\begin{cases} \rho\dfrac{d\vartheta}{dt}=\nabla\cdot\mathbf{v} \quad (\text{mass}), \\ \rho\dfrac{dY_\alpha}{dt}=\dot{W}_\alpha-\nabla\cdot\mathcal{J}_{D\alpha} \\ \rho_\alpha\dfrac{de_{\alpha,t}}{dt}=\dot{W}_{E\alpha,t}-\nabla\cdot\mathbf{q}_{\alpha,t}-e_{\alpha,t}\left(\dot{W}_\alpha-\nabla\cdot\mathcal{J}_{D\alpha}\right) \\ +\varphi_\alpha-p_\alpha\nabla\cdot\mathbf{v}+\mathcal{J}_{D\alpha}\cdot f_\alpha \qquad (\text{internal translational energy}) \\ \rho_\alpha\dfrac{de_{\alpha,\text{int}\neq t}}{dt}=\dot{W}_{E\alpha,\text{int}}-\nabla\cdot\mathbf{q}_{\alpha,\text{int}}-e_{\alpha,\text{int}}\left(\dot{W}_\alpha-\nabla\cdot\mathcal{J}_{D\alpha}\right) \\ \left(\begin{array}{l}\text{internal energy of the internal modes except for} \\ \text{translational motion}\end{array}\right) \end{cases} \qquad [1.122]$$

The entropy balance equation has the classic form:

$$\rho \frac{ds}{dt} + \nabla \cdot \mathcal{J}_S = \dot{W}_S \qquad [1.123]$$

with the entropy flux:

$$\mathcal{J}_S = \sum_{\alpha,\text{int}} \frac{\mathbf{q}_{\alpha,\text{int}}}{T_{\alpha,\text{int}}} - \frac{g_{\alpha,\text{int}}}{T_{\alpha,\text{int}}} \mathcal{J}_{D\alpha} \qquad [1.124]$$

with the chemical potentials:

$$g_{\alpha,t} = e_{\alpha,t} + p_\alpha / \rho_\alpha - T_{\alpha,t} s_{\alpha,t}, \quad g_{\alpha,\text{int}} - T_{\alpha,\text{int}} s_{\alpha,\text{int}} \qquad [1.125]$$

The resulting entropy production rate has the expression:

$$\dot{W}_S = \sum_\alpha \frac{\varphi_\alpha}{T_{\alpha,t}} - \sum_\alpha \dot{W}_\alpha \sum_{\text{int}} \frac{g_{\alpha,\text{int}}}{T_{\alpha,\text{int}}} + \sum_{\alpha,\text{int}} \frac{\dot{W}_{E\alpha,\text{int}}}{T_{\alpha,\text{int}}} + \sum_{\alpha,\text{int}} \left(\mathcal{J}_{D\alpha} \cdot \frac{f_\alpha}{T_{\alpha,t}} \right)$$
$$+ \sum_{\alpha,\text{int}} \left(\mathbf{q}_{\alpha,\text{int}} \cdot \nabla \frac{1}{T_{\alpha,\text{int}}} \right) - \sum_\alpha \left(\mathcal{J}_{D\alpha} \cdot \nabla \frac{g_{\alpha,\text{int}}}{T_{\alpha,\text{int}}} \right) \qquad [1.126]$$

1.3.2. *Application*

Here, we are interested in high-temperature flows of air in nozzles.

1.3.2.1. *Physical modeling*

1.3.2.1.1. Hypotheses

If we neglect ionization, air can be modeled as a mixture of five chemical species (N_2, O_2, NO, N and O), each of which can be treated as a perfect gas. In addition, we ignore the effects of radiation and external forces.

We shall solve mono-dimensional ("quasi-one-dimensional") equations. The viscosity and diffusion of species are not taken into account. The only irreversibilities are due to the chemical reactions and to the non-equilibria of the internal modes of energy.

The internal modes of translational, rotational and electron excitation energy are assumed to be at thermal equilibrium[5], giving us a common temperature T, while the taking into account of the vibrational non-equilibrium gives us as many temperatures as there are molecular species.

1.3.2.1.2. Differential system to be solved

The system of partial differential equations governing the flow in a nozzle is written as follows [FLA 89; FLA 90]:

$$\frac{\partial \mathbf{W}}{\partial t}+\frac{\partial \mathbf{F}}{\partial x}+\mathbf{H}=\hat{\mathbf{\Omega}} \qquad [1.127]$$

where

$$\underline{\mathbf{W}}=S\begin{Vmatrix}[\rho_\alpha]\\ \rho U\\ \rho(e+k)\\ \left[\rho_\beta e_{\beta;v}\right]\end{Vmatrix},\quad \underline{\mathbf{F}}=S\begin{Vmatrix}[\rho_\alpha U]\\ p+\rho U^2\\ pU+\rho(e+k)U\\ \left[\rho_\beta e_{\beta;v}U\right]\end{Vmatrix},$$

$$\underline{\mathbf{H}}=\begin{Vmatrix}[0_\alpha]\\ -p\dfrac{\partial S}{\partial x}\\ 0\\ \left[0_\beta\right]\end{Vmatrix},\quad \underline{\hat{\mathbf{\Omega}}}=S\begin{Vmatrix}\left[\dot{W}_\alpha\right]\\ 0\\ 0\\ \left[\dot{W}_{co-e_{\beta;v}}+e_{\beta;v}\dot{W}_\beta+\dot{W}_{d-e_{\beta;v}}\right]\end{Vmatrix}$$

where $S=S(x)$ is the section of the nozzle, ρ the density of the mixture, U the velocity, $(e+k)$ the total energy per unit mass, p the pressure of the mixture, $e_{\beta;v}$ the vibration energy of species β per unit mass, $\dot{W}_{co-e_{\beta;v}}$ the vibration energy production of species β due to non-reactive collisions per unit volume. The square brackets [ ] between double matrix bars $\| \ \|$ themselves represent column matrices with $\alpha=1,2,3,4,5;\ \beta=1,2,3$.

1.3.2.1.3. Thermodynamic model

The temperature and pressure of the mixture are written thus [FLA 89; FLA 90]:

5 This is a common hypothesis, as the characteristic times are much shorter than those of the other modes.

$$\begin{cases} T = \left[\rho(e+k) - \rho U^2/2 - \sum_{\alpha} \rho_\alpha \left(e_{\alpha;el} + h_\alpha^0 \right) - \sum_{\beta} \rho_\beta \, e_{\beta;v} \right] \Big/ \sum_{\alpha} \rho_\alpha \, c_{v;\alpha;t,r} \\ p = \sum_{\alpha} \rho_\alpha RT / \mathcal{M}_\alpha \end{cases}$$

[1.128]

where $e_{\alpha;el}$ is the energy of electron excitation per unit mass of species α, h_α^0 is the mass enthalpy of formation of species α and $c_{v;\alpha;t,r}$ is the specific heat of species α per unit mass taking account only of the translational and rotational DOFs (Appendices 2 and 4).

1.3.2.1.4. Chemical model

Here, we use the Gardiner model [GAR 84]. This model uses the following reversible reactions:

$$\begin{cases} N_2 + M & \rightleftharpoons & N + N + M \\ O_2 + M & \rightleftharpoons & O + O + M \\ NO + M & \rightleftharpoons & N + O + M \\ N_2 + O & \rightleftharpoons & NO + N \\ NO + O & \rightleftharpoons & O_2 + N \end{cases}$$

$$M = N_2, O_2, NO, N, O$$

The symbolic expression of these reactions gives us a formula similar to equation [2.11] in [PRU 12], but with reversible reactions $\rightleftarrows$ [6]:

$$\sum_{\alpha} \nu'_\alpha \mathcal{E}_\alpha \underset{k_{R;r}}{\overset{k_{D;r}}{\rightleftarrows}} \sum_{\alpha} \nu''_\alpha \mathcal{E}_\alpha \qquad [1.129]$$

where r denotes the chemical reaction, $\nu'_{\alpha;r}$ and $\nu''_{\alpha;r}$ are the stoichiometric coefficients of that reaction, $\mathcal{E}_\alpha$ represents the species α, $k_{D;r}$ is the direct

6 Note that here and as indicated in the list of symbols, the species $\mathcal{E}$ is represented by the same letter as the internal energy E of a system. However, there is a difference in style to differentiate them: the former is presented in *Monotype Corsiva* and the second in *Times New Roman*.

constant of the reaction r, and $k_{R;r}$ is the inverse constant of the reaction r (R in this case stands for Return).

The term of chemical production $\dot{W}_\alpha$ has the expression

$$\dot{W}_\alpha = \mathcal{M}_\alpha \sum_r \left(\nu''_{\alpha;r} - \nu'_{\alpha;r}\right)\left[k_{D;r}\prod_\gamma C_\gamma^{\nu'_{\gamma;r}} - k_{R;r}\prod_\gamma C_\gamma^{\nu''_{\gamma;r}}\right] \quad [1.130]$$

The direct and inverse constants of the reaction r, $k_{D;r}$ and $k_{R;r}$, obey an Arrhenius law (see [PRU 12] for example):

$$\begin{cases} k_{D;r} = B_{D;r} T^{-C_{D;r}} e^{-D_{D;r}/kT} \\ k_{R;r} = B_{R;r} T^{-C_{R;r}} e^{-D_{R;r}/kT} \end{cases} \quad [1.131]$$

where k is the Boltzmann constant and the different coefficients are constants (see Appendix 2).

1.3.2.1.5. Vibration model

In equation [4.127] in [VIN 86], $\dot{W}_{co-e_{\beta;v}}$ contains only the translation/vibration interaction:

$$\dot{W}_{co-e_{\beta;v}} = \rho_\beta \frac{e_{\beta;v}(T) - e_{\beta;v}}{\tau_{\beta;v}}, \quad \frac{1}{\tau_{\beta;v}} = \sum_\gamma \frac{X_\gamma}{\tau_{v;\beta;\gamma}} \quad [1.132]$$

where X_γ is the molar fraction of species γ $(\gamma = 1,\ldots,5)$ and the characteristic times $\tau_{v;\beta;\gamma}$ are given by Millikan & White [MIL 63].

The vibration temperature $T_{\beta;v}$ of the diatomic molecular species β is derived from the computation variables ρ_β and $\rho_\beta e_{\beta;v}$ by the implicit equation: $e_{\beta;v}(T_{\beta;v}) = \rho_\beta e_{\beta;v} / \rho_\beta$, where $e_{\beta;v}(T_{\beta;v})$ is an analytical function of $T_{\beta;v}$ taking account of the anharmonicity and the rotation/vibration couples.

1.3.2.1.6. Rotation/vibration coupling model

This coupling refers only to the influence of vibrational relaxation on the chemical kinetics. Vibrational relaxation intervenes directly in the reaction constants. Thus we have, only with dissociations and in the forward direction, [PAR 88]:

$$k_{D;r} = B_{D;r}\, T_{\beta(r);a}\, e^{-D_{\beta(r)}/k\, T_{\beta(r);a}} \qquad [1.133]$$

where $T_{\beta(r);a}$ is a midway temperature between the translational temperature and vibrational temperature [PAR 88; SHA 88]: $T_{\beta(r);a} = T^{s}\left(T_{\beta(r);v}\right)^{1-s}$ taking account of the temperature of the mixture T and the vibrational temperature $T_{\beta(r);v}$ of the diatomic species β (the subscript $\beta(r)$ corresponding to the molecule β dissociated in the reaction r) where s is a real number between 0.5 and 0.7. The influence of the chemical kinetics on the vibrational relaxation is non-existent: $\dot{W}_{d-e_{\beta;v}} = 0$.

1.3.2.2. *Evolution of the entropy of the mixture*

1.3.2.2.1. Entropy production

Given the hypotheses from section 1.3.2.1, the rate of entropy production [PRU 88; FLA 93] is reduced to:

$$\dot{W}_S = \sum_{\beta} \dot{W}_{E\beta;v}\left(\frac{1}{T_{\beta;v}} - \frac{1}{T}\right) - \sum_{\alpha} \dot{W}_{\alpha}\frac{g_{\alpha;t}}{T} - \sum_{\beta} \dot{W}_{\beta}\frac{g_{\beta;v}}{T_{\beta;v}} \qquad [1.134]$$

and the entropy flux is zero: $\vec{\boldsymbol{J}}_S = \vec{\boldsymbol{0}}$. It follows from this that the rate of entropy production can be directly deduced from the calculation of the entropy and its material derivative. In a quasi-one-dimensional steady flow, the expression of this is

$$\dot{W}_S = \rho U\, ds/dx \qquad [1.135]$$

Thus, here, we shall limit ourselves to monitoring the evolution of the entropy along the axis of the nozzle.

1.3.2.2.2. Expression of entropy

The expression of the entropy of the mixture [PRU 91a] is obtained by taking account of the five chemical species, the contribution of the electronic energy in addition to the translational and rotational energy (the three modes being in equilibrium in the mixture), and the existence of three vibrators that cannot be described as simple harmonic oscillators. We can write the expressions of the partition functions of each degree of species α at standard pressure p^0. From this, we shall deduce the *standard* entropies by simple differentiation (see Appendices 2 and 4).

The volumetric entropy of the mixture has the expression

$$\rho s = \sum_{\alpha} \rho_\alpha \frac{e_{\alpha;t} + e_{\alpha;r} + e_{\alpha;el}}{T} + \sum_{\beta} \frac{e_{\beta;v}}{T_{\beta;v}} + \sum_{\alpha} \rho_\alpha \frac{R}{\mathcal{M}_\alpha}$$
$$+ \sum_{\alpha} \rho_\alpha \frac{R}{\mathcal{M}_\alpha} \left[\ln\left(Q_\alpha^0 k T^{3/2} Q_{\alpha;r} Q_{\alpha;el}\right) + \ln\left(\frac{\mathcal{M}_\alpha}{\rho_\alpha R}\right) \right] + \sum_{\beta} \rho_\beta \frac{R}{\mathcal{M}_\beta} \ln\left(Q_{\beta;v}\right) \qquad [1.136]$$

with $Q_\alpha^0 = \left(\frac{2\pi k^2 \mathcal{M}_\alpha}{R h^2} \right)^{3/2}$ where h is Planck's constant.

1.3.2.3. *Numerical approach and configuration of the nozzles in question*

The ten equations in system [1.127] are solved in their conservative form by a *totally implicit de-centered method* [FLA 89; FLA 90] developed by Harten and Yee [YEE 85; YEE 87; HOL 89] where in particular, aerodynamics, chemistry and vibration are strongly coupled.

This method enables us to calculate the flow through the entirety of the nozzle, including in difficult equilibrium areas and of subsonic velocity.

The method is applied to the calculation of *two hypersonic nozzles* reaching Mach numbers of six and twelve. The fluid entering the nozzle is assumed to be at thermochemical equilibrium, at the pressure and temperature of the reservoir.

The first nozzle examined is an R6-type nozzle, and was the subject of an in-depth study in previous works [FLA 89; FLA 90]. The generative conditions are $p = 50$ bars and $T = 4800$ K. The second hypersonic nozzle used is an F4 nozzle, with a cone of 11.3° for the divergent section and reservoir conditions of $p_i = 2000$ bars and $T_i = 8508$ K. The throat is situated at 9.4 mm from the opening for a cross-section of 1 cm^2. The calculations were carried out up to $x = 2$ m, which corresponds to two thirds of the length of the divergent cross-section of the nozzle.

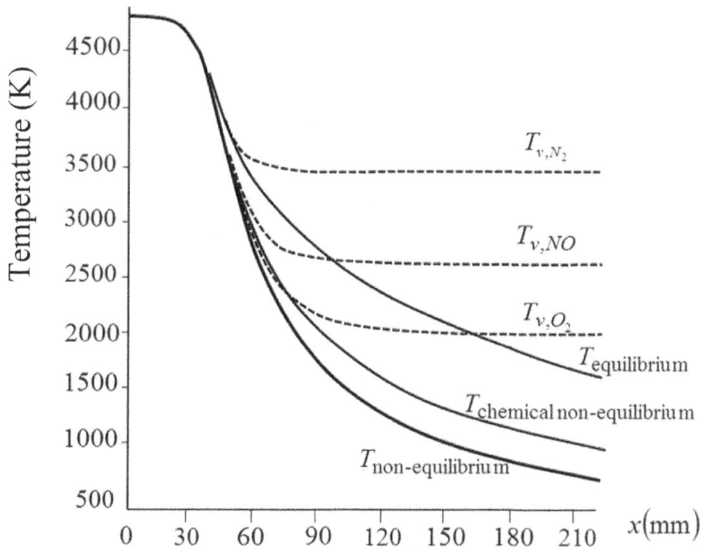

Figure 1.9. *Temperature distributions in the R6 nozzle of the ONERA (see Flament & Prud'homme [FLA 93])*

1.3.2.4. *Results*

The calculations were performed on the CRAY-XMP supercomputer, with a uniform mesh of 225 points for the R6 nozzle and a mesh defined by geometric progressions on both sides of the throat for the F4 nozzle. The computation time for the convergence obtained after a fivefold decrease of the residuals are 20 minutes for the R6 nozzle and 1 hour for the F4 nozzle, as the latter has far more difficult generative conditions.

Many studies have been carried out on these nozzles [FLA 89; FLA 90]. The results presented here relate only to the R6.

1.3.2.4.1. Comparison with equilibrium

Figure 1.9 shows the distribution of temperatures in the case of the R6 nozzle (for the F4 nozzle, see [FLA 93]). We can see that in the upstream part, the flow is at thermal equilibrium and the computation method has indeed been able to simulate the phenomenon.

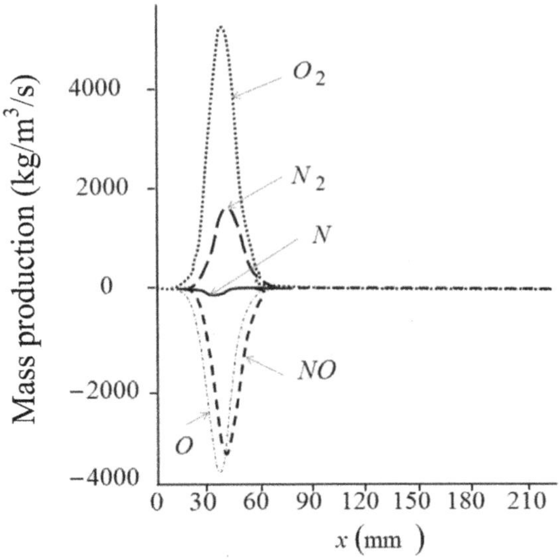

Figure 1.10. *Distribution of the mass production of species in the R6 nozzle of the ONERA (see [FLA 93])*

1.3.2.4.2. Production of species

Figure 1.10 shows the distributions of mass productions of the different species in the mixture. For each species, we see three distinct zones: an initial part with no production at all, delimiting the chemical equilibrium zone; a second part with significant production determining the non-equilibrium zone, and a third part with negligible production, determining the frozen zone. It should be noted that the code captures the three zones of flow along the nozzle in a single calculation.

1.3.2.4.3. Entropy production of the mixture

In the entropy distribution for the F4 nozzle, similarly as for the mass production, we can distinguish three zones: an initial zone with constant entropy (equilibrium), a second zone with significantly increasing entropy (non-equilibrium) and a third zone with slightly decreasing entropy (frozen part). We also observe (see [FLA 93]), as predicted by equation [1.135], a positive variation in entropy along the nozzle. The curve shows a slight disturbance in the vicinity of the throat, indicating an inadequacy of the digital treatment of the physical problem with source terms in the sonic zone.

1.4. Conclusion about flows in nozzles

The restrictive hypothesis of the ideal mixture enables us to deal with fairly varied non-equilibrium situations in the gases: chemical non-equilibria, internal modes, transfer phenomena. Particular attention has been paid to the case of a mixture with multiple translational and vibrational temperatures. Translational non-equilibrium arises when the species present have very different molar masses (e.g. hydrogen and nitrogen or oxygen) and occurs very frequently in plasmas where the electron is the lightest species.

When the molar masses of the species are not too different and ionization is very slight, we can assume that there is only one translational temperature. We can accept that this is true with flows of air at high velocity, where only the chemical non-equilibria and the vibrational relaxation, which are generally coupled, remain.

In this chapter, we have limited ourselves to a mono-dimensional hypothesis for nozzle flows. We also need to deal with two-dimensional reactive flows in transonic nozzles, by using the asymptotic expansion method, and there is no shortage of numerical codes of two- or three-dimensional reactive flow in nozzles, which also take account of turbulence [SAI 12].

These problems, and others involving boundary layers, expansion waves and shockwaves in multi-species flow, presented in Chapter 1 of [PRU 12] in a simple fluid, are discussed in detail in many books (e.g. [PRU 10] with regard to 2D nozzles, the laminar and turbulent boundary layer over a fixed flat plate, or over a rotating disc with diffusion of species).

Remember finally that Chapter 3 is entirely given over to laminar and turbulent reactive flows found in combustion.

Chapter 2

Chemical Reactors

Chemical engineering (sometimes called process engineering) is a very important domain, with its own specific analysis and solution methods [VIL 82; TRA 04; MOR 10].[1]

Flows, which experience complex chemical reactions, are to be found in chemical reactors, used for the synthesis of new substances, on a laboratorial or industrial scale. They play a very significant part in petrochemistry, particularly in refining but also in microelectronics. Microscopic reactors are designed for biomedical installations or for *in situ* production of highly toxic or dangerous compounds. The elaboration of less pollutant products requires the development of new processes, e.g. for hydrodesulfurization in oil-refining.[2]

Often, the conventional methods for analysis and modeling of aerodynamics and combustion are not the most appropriate for solving problems relating to chemical reactors. We tend to use deterministic and probabilistic population balance equations such as those given in Table 2.1 in [PRU 12]. These are too refined in the case of

---

1 The author wishes to thank Patrick DA COSTA for his attentive proofreading of this chapter, and for the information he provided about industrial applications.

2 Sulfur oxides are responsible for pollution smog over cities, and for acid rain which can cause deforestation, acidification of water and soil, and damage to building façades. Respiratory problems in humans – particularly in sensitive people and young children, can also be caused by these pollutants. In order to produce cleaner fuels, the sulfur contained in crude oil is removed during refining by a hydrotreatment process. Hydrogen reacts with the hydrocarbon molecules containing sulfur atoms to form hydrogen sulfide (H_2S). The hydrogen sulfide generated is then oxidized in sulfur removal units, sometimes using oxygen. The sulfur thus recovered can be valorized in the chemical industry (fertilizer manufacture) to make batteries or rubber. See http://www.airliquide.com/en/fuel-desulfurization.html.

chemical reactors, for which precise knowledge of the field of velocities is not important. In addition, they are sometimes difficult to use for interpreting experimental results.

In the first section of this chapter, we demonstrate the differences between ideal and real reactors. Then, in section 2.2, we outline the behavior of a well-known kind of ideal reactor: a perfectly stirred homogenous reactor in a steady or unsteady regime.

Paradoxically, there are certain similarities between the properties of stability of this type of chemical reactor and those of the reaction zone in a premixed flame. We shall touch on this question when we discuss turbulent combustion in Chapter 3.

The tubular reactor is presented in section 2.3.

Residence time distribution is studied in section 2.4. The basic equations for residence time distribution are established and then applied to ideal reactors: to a perfectly stirred homogeneous reactor in the permanent regime, to a plug flow reactor, to a Poiseuille flow and finally to real reactors.

2.1. Ideal reactors, real reactors, balance equations

2.1.1. *Ideal chemical reactors*

A chemical reactor, in the broadest sense, is an area of space where a chemical transformation can take place. This definition covers a huge variety of situations: industrial chemistry, thrusters for aircraft, rockets, chemical engines, etc.

The configuration of a reactor depends on many factors relating to the chemical nature of the transformation taking place: a homogeneous or heterogeneous reaction, the difference from equilibrium, the presence of catalysts, turbulence, the thermodynamic state (pressure, temperature, etc.).

In order to study real reactors, we need to develop experimental techniques and means of theoretical investigation. The residence time of the chemical species in a reactor is an important notion, and examining the residence time distribution often gives us a picture of the process in all its complexity. The study of extreme cases, i.e. *ideal reactors* (Figure 2.1), is advantageous because they are simpler and because they often offer a good approximation of real cases.

We can classify reactors on the basis of the number of phases involved, whether or not there is stirring and finally the operating mode (a continuous, semi-continuous or discontinuous process).[3]

It is often an impossibility to solve the local balance equations without making any simplifying hypotheses. We generally use global balances which take account of the main characteristics of the reactors. In certain cases, one could, for instance, neglect the transfer phenomena (heat conduction, diffusion, viscosity). Sometimes – as is the case with isothermal reactors – only the species balance equation will be used [BAR 70].

For a so-called "perfectly stirred" (or "perfectly mixed") ideal reactor, we can distinguish only the input conditions and the (unique) conditions in the reactor itself, which are also the output conditions. This hypothesis is no longer valid for a "plug flow" reactor, where we suppose the flow to take place in successive stages, leading to the output. Finally, in most of the cases envisaged in this chapter, the effects of the variation in pressure and viscosity will be overlooked, so the momentum equation will not be used.

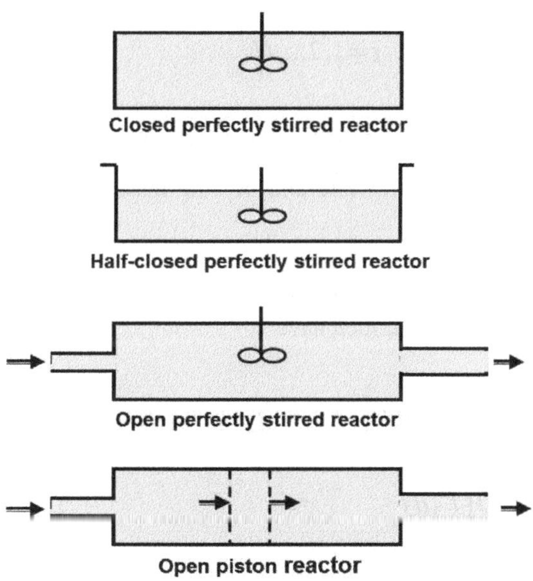

Figure 2.1. *Various types of ideal reactors*

3 Section 6.3, *Chemical reactors*, in the *Enciclopedia Trecanni* [ENC 08], gives a table showing nine types of chemical reactors, with notes about the characteristics of their use.

We shall examine the permanent regime and sometimes the transitory regime and the stability of the points of operation of some ideal reactors.

2.1.2. *Balance equations for chemical reactors*

In this section, we shall begin by presenting the global deterministic balances, which will then enable us to study certain so-called "ideal" reactors. Turbulent reactive flows (see Chapter 3) or real chemical reactors will require the establishment of probabilistic balance equations, which are presented later on.

2.1.2.1. *Global deterministic balance*

In principle, we need only write the integral form of the balance equations established in Chapter 2 of [PRU 12]. We generally take account of the fact that the opening and exit of a reactor comprise pipes, along the length of which the properties of the fluids are relatively uniform. In addition, the reactor or element of the reactor in question has non-deformable walls. The mass balance of the species *j* will therefore be:

$$\frac{dm_j}{dt} = \dot{m}_{jo} - \dot{m}_{je} + R_j,\ j = 1,2,\dots N \qquad [2.1]$$

where m_j is the mass of species *j* contained in the reactor, $\dot{m}_{jo}$ is the opening mass flowrate, $\dot{m}_{je}$ the exit mass flowrate and R_j the mass produced by the chemical reaction.

Let us apply the first law of thermodynamics to the open reactor in Figure 2.2 between times *t* and $t + dt$. We then have

$$\left(E + dE + \sum_j \dot{m}_{je}\, e_{je}\, dt\right) - \left(E + \sum_j \dot{m}_{jo}\, e_{jo}\, dt\right) = \frac{p_o}{\rho_o}\dot{m}_o\, dt - \frac{p_e}{\rho_e}\dot{m}_e\, dt - \dot{Q}(T)\, dt \qquad [2.2]$$

where $\dot{Q}(T)$ denotes the quantity of heat lost per unit time through the walls. From this we deduce:

$$\frac{dE}{dt} = \sum_j \dot{m}_{jo} h_{jo} - \sum_j \dot{m}_{je} h_{je} - \dot{Q}(T) \qquad [2.3]$$

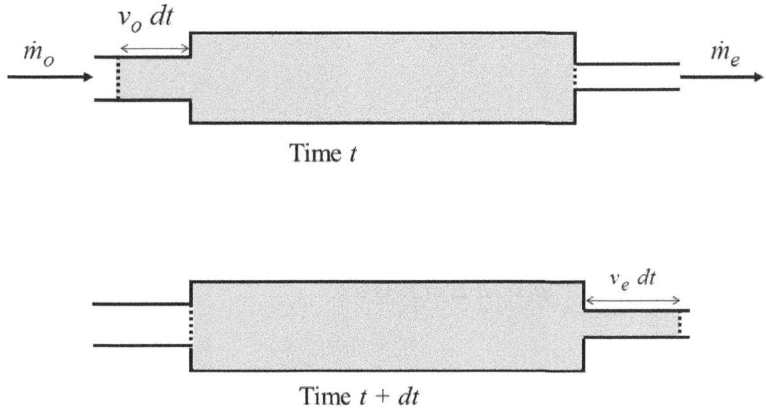

Figure 2.2. *Mass balance in an open reactor*

These (N+1) equations are sufficient to solve numerous problems. Thus, for a perfectly mixed ideal reactor, we accept that the specific or intensive quantities are identical in the exit section of the reactor and at any point inside the reactor. This implies a highly turbulent fluid mixture, and leads us to a simple formulation.

2.1.2.2. *Probabilistic population balance*

The probabilistic population balance method [DAN 51; VIL 82] is applicable to chemical reactors. We accept that an entity of fluid, in position $\vec{\mathbf{x}}$ at time t, is characterized by a certain number of values $\zeta_1, \zeta_2, \ldots$ forming a vector $\vec{\zeta}$. The number of entities contained in the element of volume $\mathbf{d\vec{x}\,d\vec{\zeta}}$ in the phase space thus defined is, at time t:

$$\psi\left(\vec{\mathbf{x}}, \vec{\zeta}, t\right) d\,\vec{\mathbf{x}}\, d\,\vec{\zeta} \qquad [2.4]$$

and the total number of entities contained in a control volume $\left(\mathcal{V}\right)$ is then:

$$N = \int_{\mathcal{V}} \psi\left(\vec{\mathbf{x}}, \vec{\zeta}, t\right) d\,\vec{\mathbf{x}}\, d\vec{\zeta} \qquad [2.5]$$

and here, the notation N does not correspond to the number of species in the mixture.

The possibility that an entity with volume $\mathcal{V}$ at position $\vec{\mathbf{x}}$ at time t will be in the state defined by $\vec{\zeta}$ is:

$$f = \psi / N \quad [2.6]$$

The population balance is written:

$$\frac{\partial}{\partial t}\left(\int_{\mathcal{V}} \psi \, d\mathcal{V}\right) + \int_{\partial\mathcal{V}} \psi \vec{\mathbf{V}} \cdot d\vec{\mathbf{\Sigma}} = \int_{\mathcal{V}} G \, d\mathcal{V} \quad [2.7]$$

where $(\partial\mathcal{V})$ is the fixed closed surface delimiting the volume $(\mathcal{V})$, and $d\vec{\mathbf{\Sigma}}$ is the element of area exiting in a direction normal to $(\partial\mathcal{V})$. The vector $\vec{\mathbf{V}}$ is the velocity vector of the phase space. The rate of production of entities per unit volume of that space is defined by G.

If we set $\vec{\mathbf{w}} = d\vec{\zeta}/dt$, we have the local equation:

$$\frac{\partial \psi}{\partial t} + \vec{\nabla} \cdot (\psi \vec{\mathbf{v}}) + \sum_i \frac{\partial (\psi w_i)}{\partial \zeta_i} = G \quad [2.8]$$

By integrating this equation into the physical volume (υ) of an open reactor with opening and exit sections denoted by (o) and (e), with opening volume flowrate $\dot{q}_o$ and exit volume flowrate $\dot{q}_e$, we obtain

$$\frac{1}{\upsilon}\frac{\partial (\overline{\psi} \upsilon)}{\partial t} + \frac{\dot{q}_e \overline{\psi}_e - \dot{q}_o \overline{\psi}_o}{\upsilon} + \sum_i \frac{\partial (\overline{\psi} w_i)}{\partial \zeta_i} = \overline{G} \quad [2.9]$$

In this equation, we see the involvement of the average volumetric values:

$$\overline{\psi} = \frac{1}{\upsilon}\int_{\upsilon} \psi \, d\mathcal{V}, \quad \overline{G} = \frac{1}{\upsilon}\int_{\upsilon} G \, d\mathcal{V} \quad [2.10]$$

and the average volumetric flowrates:

$$\bar{\psi}_o = \frac{1}{\dot{q}_o}\int_{Ao} \psi \vec{\mathbf{v}} \cdot d\vec{\mathbf{A}},\ \bar{\psi}_e = \frac{1}{\dot{q}_e}\int_{A_e} \psi \vec{\mathbf{v}} \cdot d\vec{\mathbf{A}} \qquad [2.11]$$

This balance equation can be applied, for instance, to a crystallizer (Figure 2.3). The input solution is homogeneous and the crystals appear in the reactor with a rate of nucleation r_N. Their dimension L increases with a certain rate $w = dL/dt = r_0 L^b$, where r_0 and b are constants.

The reactor is assumed to be perfectly stirred. For $b = 0$, in the steady regime, we find a law which is remarkably well borne out by experience, giving the ratio $W(L)$ of the mass $m(L)$ of crystals whose size is less than a given L, to the total crystallized mass m_T (Figure 2.4). The deviation $W'(L)$ is:

$$W'(L) = \frac{m'(L)}{m_T} = \frac{1}{6L_0}\left(\frac{L}{L_0}\right)^3 e^{-L/L_0} \qquad [2.12]$$

where $L_0 = \upsilon r_0/\dot{q}$, $\dot{q} = \dot{q}_o = \dot{q}_e$ is the volumetric flowrate.

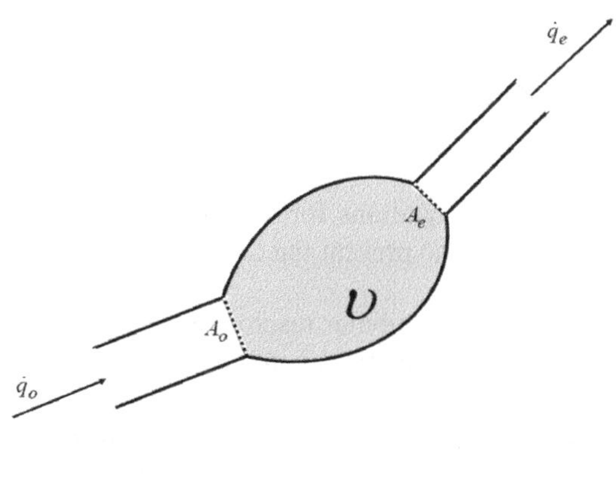

Figure 2.3. *The reagents enter into the crystallizer reactor with the flowrate $\dot{q}_o$, and the products exit with the flowrate $\dot{q}_e$*

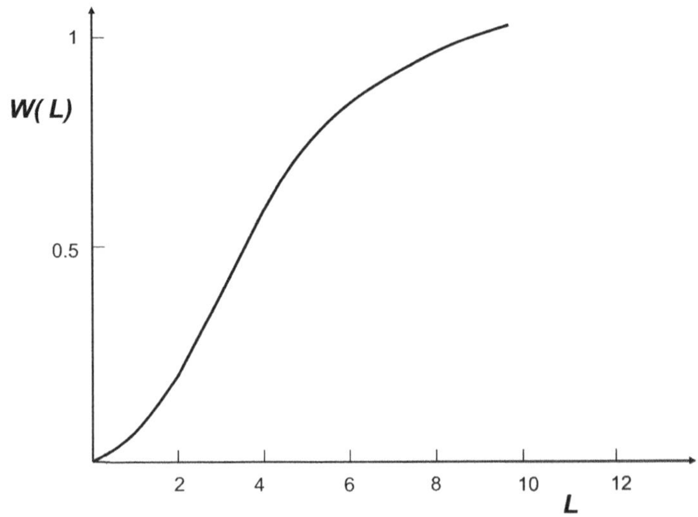

Figure 2.4. *$W(L)$ fraction of mass of crystals of size less than a given L*

2.2. Perfectly mixed homogeneous chemical reactors

In this section, we are going to look at perfectly mixed reactors. These reactors are highly flexible in terms of their implementation, and ensure that the reagents are well mixed. They are used for organic reactions in fine chemistry or in pharmacology, melanin production, the reactions of esterification, saponification, etc. (see *Enciclopedia Treccani* [ENC 08] cited above in section 2.1.1).

Having discussed the basic equations for such reactors, we look at their steady regimes and their stability. We also present the conditions for their stabilization.

A comparable study is done for tubular reactors in section 2.3. The performances of these reactors mean they are sometime favored over stirred reactors.

These notions of chemical reactors do not lend themselves to the study of combustion waves. We examine combustion waves in Chapter 3, where we set about solving the general balance equations of reactive flows involving all the transfer phenomena.

2.2.1. *Equations for a perfectly stirred homogeneous chemical reactor*

With a perfectly stirred homogeneous open reactor, we use the global balance equations in section 2.1.2:

$$\frac{dm_j}{dt} = \dot{m}_{jo} - \dot{m}_{je} + R_j,\ j = 1,2,...N \qquad [2.13]$$

$$\frac{dE}{dt} = \sum_j \dot{m}_{jo} h_{jo} - \sum_j \dot{m}_{je} h_{je} - \dot{Q}(T) \qquad [2.14]$$

By the term "perfectly stirred", we understand that the thermodynamic and chemical parameters are the same at all points in the reactor, once the mixture has passed the opening point. In particular, $T_e = T$, $h_{je} = h_j, C_{je} = C_j$.

In the case in point, we accept that the pressure p is approximately constant, so that, if $\mathcal{V}_r$ is the volume of the reactor, $dE/dT = dH/dT$, when $H = \mathcal{V}_r \sum_j \rho_j h_j$.

Finally, the mass flowrates $\dot{m}_j$ are assumed to be proportional to the densities, such that, if $\dot{q}$ is the volumetric flowrate:

$$\dot{m}_j = \rho_j \dot{q} \qquad [2.15]$$

If we set

$$\tau = \mathcal{V}_r / \dot{q} \qquad [2.16]$$

the passage time, equal to an average residence time, we have

$$\begin{cases} \tau\, d\rho_j / dT = \rho_{jo} - \rho_j + \tau \dot{W}_j \\ \tau\, d\left(\sum_j \rho_j h_j \right) \Big/ dt = \sum_j \left(\rho_{jo} h_{jo} - \rho_j h_j \right) - \dot{Q}/\dot{q} \end{cases} \qquad [2.17]$$

in view of the fact that $R_j = \mathcal{V}_r \dot{W}_j$.

If only one chemical reaction occurs,

$$\dot{W}_j = \nu_j \mathcal{M}_j \dot{\zeta} \qquad [2.18]$$

where $\dot{\zeta}$ is the reaction rate in moles per unit time. We also have

$$d\rho_j = \nu_j \mathcal{M}_j d\zeta \qquad [2.19]$$

where ζ is the progress variable of the reaction such that

$$\rho_j = \rho_{jo} \nu_j \mathcal{M}_j \zeta \qquad [2.20]$$

The species balance equation is therefore written as

$$\tau d\zeta/dt = -\zeta + \tau\dot{\zeta} \qquad [2.21]$$

For the energy equation, we take account of

$$h_j = \left(q_f^0\right)_j + \int_{T^0}^{T} c_{p,j}\, dT \qquad [2.22]$$

where $\left(q_f^0\right)_j$ and $c_{p,j}$ are respectively the heat of formation and the specific heat at constant pressure per unit mass of the species j; the result of this is that

$$d\left(\sum_j \rho_j h_j\right)\Big/dt = \sum_j \left(h_j d\rho_j/dt + \rho_j c_{p,j}\, dT/dt\right) \qquad [2.23]$$

We have

$$\begin{aligned}\sum_j h_j d\rho_j/dt &= \sum_j h_j\left[\left(\rho_{jo} - \rho_j\right)/\tau + \dot{W}_j\right] \\ &= \sum_j h_j\left[\left(\rho_{jo} - \rho_j\right)/\tau + \nu_j \mathcal{M}_j \dot{\zeta}\right] \\ &= \sum_j h_j\left(\rho_{jo} - \rho_j\right)/\tau + \dot{\zeta}\sum_j \nu_j \mathcal{M}_j h_j\end{aligned} \qquad [2.24]$$

If we use the notation ΔH for the average molar enthalpy of the reaction:

$$\Delta H = \sum_j \nu_j \mathcal{M}_j h_j \tag{2.25}$$

and set

$$\rho c_p = \sum_j \rho_j c_{p,j} \tag{2.26}$$

equation [2.24] becomes

$$\sum_j (\rho_{jo} - \rho_j) h_j - \tau \Delta H \dot{\zeta} + \rho c_p \tau \, dT/dt = \sum_j (\rho_{jo} h_{jo} - \rho_j h_j) - \dot{Q}/\dot{q} \tag{2.27}$$

Simplifying this equation and setting

$$\sum_j \rho_{jo} (h_{jo} - h_j) = \sum_j \rho_{jo} \int_T^{T_o} c_{p,j} \, dT \cong \rho c_p (T_o - T) \tag{2.28}$$

we have

$$\tau \, dT/dt = T_o - T + \Delta^* \tau \dot{\zeta} - Q^* \tag{2.29}$$

where

$$\Delta^* = \Delta H / \rho c_p, \; Q^* = \dot{Q} / \rho c_p \dot{q} \tag{2.30}$$

Hence, the system defining the reactor's behavior is

$$\begin{cases} \tau \, d\zeta/dt = -\zeta + \tau \dot{\zeta}, \\ \tau \, dT/dt = T_o - T + \Delta^* \tau \dot{\zeta} - Q^*. \end{cases} \tag{2.31}$$

Let us examine the case of a first-order irreversible reaction:

$$A \xrightarrow{k(T)} B \tag{2.32}$$

where

$$k(T) = k_0 \, e^{-Ta/T} \qquad [2.33]$$

We have

$$\dot{\zeta} = k_0 e^{-T_a/T} C_A = k_0 e^{-T_a/T} (C_{A0} - \zeta) \qquad [2.34]$$

and, by setting $X = \zeta / C_{A0}$, we find

$$\dot{\zeta} = C_{A0} \, k_0 e^{-T_a/T} (1 - X) \qquad [2.35]$$

Assume that the heat exchange law is

$$Q^* = K(T - T_0) \qquad [2.36]$$

where T_0 is the external temperature, and set

$$\begin{cases} f(X,T) = k_0 e^{-T_a/T} (1 - X) \\ T^* = \Delta^* C_{A0} \end{cases} \qquad [2.37]$$

The balance equations are of the form

$$\begin{cases} \tau \, dX/dt = -X + \tau f(X,T), \\ \tau \, dT/dt = T_0 - T + \tau T^* f(X,T) - K(T - T_0) \end{cases} \qquad [2.38]$$

They are nonlinear because of the term of chemical kinetics. Thus, in particular, we are going to give the steady operating points obtained by eliminating the terms on the left-hand side of equation [2.38]. Then, we shall carry out linearization in the vicinity of these steady operating points, which will enable us to solve the equations analytically and demonstrate unstable behavior.

2.2.2. *Steady regimes in perfectly stirred homogeneous chemical reactors*

The steady regime (subscript s, Figure 2.5) is characterized by the following two equations in the plane (T_s, X_s):

$$X_S = \tau k_0 \, e^{-T_a/T_s} \Big/ \left(1 + \tau k_0 \, e^{-T_a/T_s}\right) \tag{2.39}$$

$$X_S = (K+1)\, T_s / T^* - (T_0 + K T_s)/T^* \tag{2.40}$$

The solutions correspond to the intersections of the curves representative of equations [2.39] and [2.40] (Figure 2.6).

We can also represent these solutions in the plane (X_S, f_S), with

$$f_S = f(X_s, T_s) = k_0 (1 - X_s) e^{-T_a/T_s} \tag{2.41}$$

We then obtain the following two equations:

$$\begin{cases} f_s = X_s/\tau, \\ f_s = k_0 (1 - X_s) e^{-T_a (1+K)/(T_0 + K T_0 + T^* X_s)} \end{cases} \tag{2.42}$$

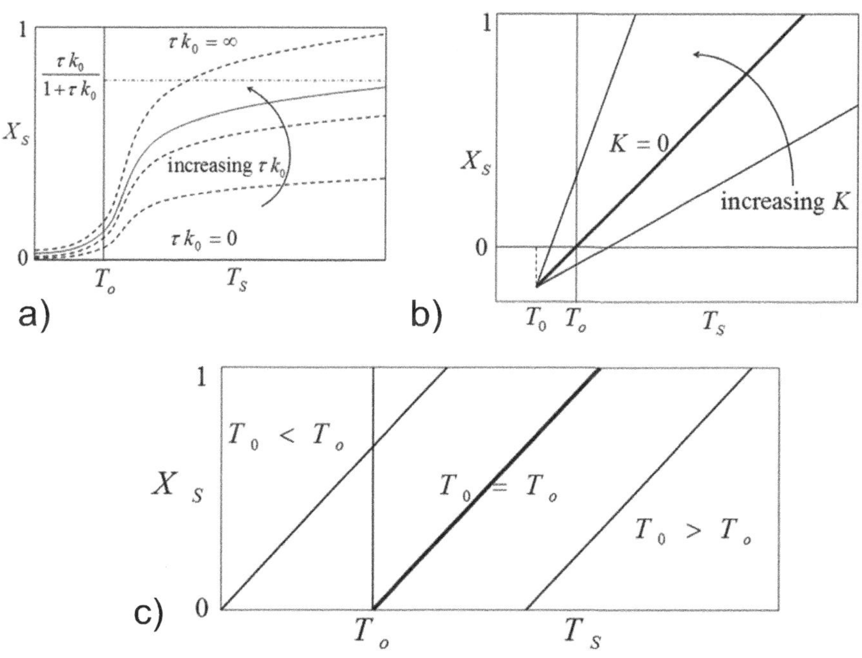

Figure 2.5. *Steady modes (equations [2.39], [2.40]). a) Influence of τk_0, which is similar to a Damhöhler number $D_a = \tau/\tau_{chem}$. With a given T_s, the larger the value of τk_0 (rapid reaction or slow flow), the larger the value of X_s.*
b) Influence of heat exchange coefficient K with given T_0 and T (here $T_0 < T_o$).
c) Influence of the temperature of the exchanger at a given K

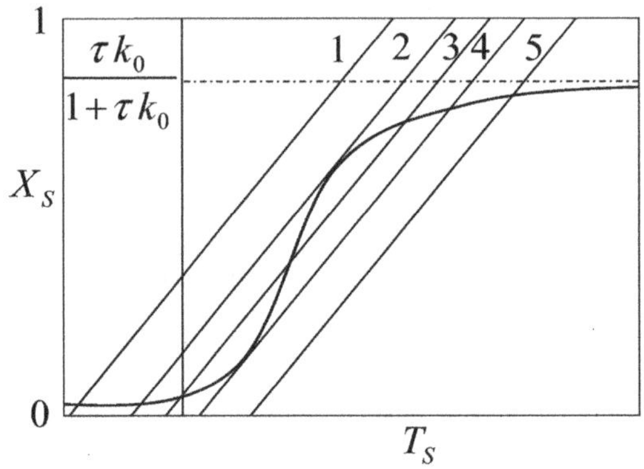

Figure 2.6. *Solutions for T_a, with given values of τk_0 and K and increasing values of T_0*

Here, the influence of the passage time is evidenced by the gradient $1/\tau$ of the line passing through the origin (Figure 2.7). We have one or three operating points, or in exceptional cases, two. The points A_i correspond to a very low reaction rate: it is advantageous to obtain as large a value of X_s as possible with a fairly short passage time τ. The reaction can be triggered by decreasing the flowrate (and therefore $1/\tau$) or increasing the temperature of the exchanger.

Thus, we have illustrated the possible operating points of the perfectly stirred reactor with a single 1st-order chemical reaction. The properties found are fairly general and are usable, as stated at the beginning of this chapter, even in combustion for the reactive zone of a premixed flame with high activation energy (see Chapter 3).

We saw earlier that it was possible to determine the operating point by adjusting the passage time and the temperature of the exchanger. We shall see in the next section that it is possible to control a reactor using signals of temperature and concentration and thereby affect the stability of the operating points.

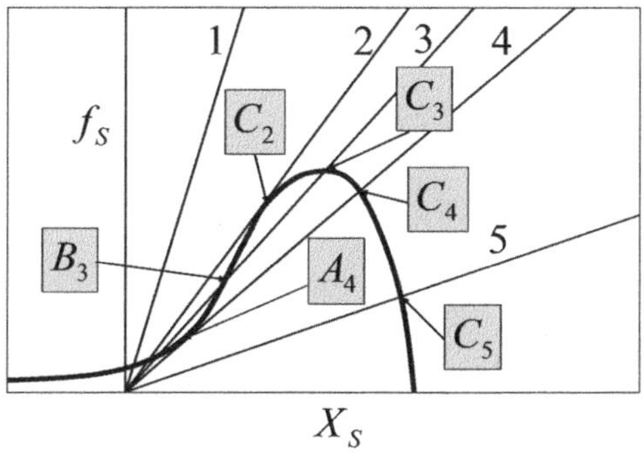

Figure 2.7. *Steady solutions in the plane* $\left(X_S, f_S\right)$

2.2.3. *Stability of operating points in the perfectly stirred homogeneous chemical reactor*

It is important to know the stability of the operating points [ARI 58]. Therefore, we look at the evolution over time of the differences from the steady regime:

$$x_1 = \left(X - X_s\right)/\left(1 - X_s\right),\ x_2 = \left(T - T_s\right)/T_s \qquad [2.43]$$

We obtain

$$\begin{cases} dx_1/d\theta = -x_1 - a_1\left\{1 - \left(1 - x_1\right)\exp\left[a_2 x_2/\left(1 + x_2\right)\right]\right\} \\ dx_2/d\theta = -a_3 x_2 - a_4\left\{1 - \left(1 - x_1\right)\exp\left[a_2 x_2/\left(1 + x_2\right)\right]\right\} \end{cases} \qquad [2.44]$$

where

$$\begin{cases} \theta = t/\tau, a_1 = \tau k_0\, e^{-T_a/T_s}, a_2 = T_a/T_s \\ a_3 = 1 + K, a_4 = T^*\left(1 - X_s\right)a_1/T_s \end{cases} \qquad [2.45]$$

After linearization of the right-hand sides of these equations, we find

$$\begin{cases} dx_1/d\theta = -\left(1 + a_1\right)x_1 + a_1 a_2 x_2 \\ dx_2/d\theta = -a_4 x_1 + \left(a_2 a_4 - a_3\right)x_2 \end{cases} \qquad [2.46]$$

The equation characteristic of the system is

$$\lambda^2 - S\lambda + P = 0 \qquad [2.47]$$

where

$$\begin{cases} S = -(1 + a_1 + a_3 - a_2 a_4) \\ P = (1 + a_1) a_3 - a_2 a_4 \end{cases} \qquad [2.48]$$

The roots are real and distinct if

$$\Delta = (1 + a_1 - a_3)^2 + 2a_2 a_4 (1 - a_1 - a_3) + a_2{}^2 a_4{}^2 > 0 \qquad [2.49]$$

They are conjugate imaginary when $\Delta < 0$, and we find a double root when $\Delta = 0$.

1st case: non-null real roots ($\Delta > 0$)

There is stability ($\lambda_1 < 0, \lambda_2 < 0$) on condition that P is positive and S negative. Any slight deviation from the steady operating point tends to resorb in accordance with an exponential law. If only one of the roots is negative or if both are positive, we have instability.

2nd case: conjugate imaginary roots ($\Delta < 0$)

In this scenario, in all cases, $P > O$. The real part of each root λ is equal to $S/2$. Thus, stability will be assured if $S < 0$ and the solution will then be damped oscillatory in nature. If S is positive, there is instability. If S is zero, the solution is non-damped oscillatory.

3rd case: double root ($\Delta = 0$)

The solution is stable if $S < 0$.

4th case: zero root (P = 0)

There is asymptotic stability. The deviation tends to stabilize toward a non-zero value when $\theta \to \infty$.

In summary, the stability of the linearized system is only assured when $P > 0$ and $S < 0$.

One of these conditions can be represented graphically. Indeed, in the plane (T_s, X_s), p_1 is the slope of the curve $X_S = \tau k_0 e^{-T_a/T_s} / \left(1 + \tau k_0 e^{-T_a/T_s}\right)$ and p_2 is that of the straight line $X_S = (K+1)\, T_s / T^* - (T_o + K T_0)/T^*$. By differentiation, we find

$$1 - p_1/p_2 = \left[(1+a_1)a_3 - a_2 a_4\right] / (1+a_1)a_3 = P/(1+a_1)a_3 \qquad [2.50]$$

Thus, P will be positive if the slope of the straight line is greater than that of the curve, and P will be negative otherwise. Hence, we find a sufficient condition for instability with $p_2 < p_1$, which occurs only between points A and B. The arc AB is therefore a zone of unstable steady regimes for small disturbances.

Only by looking at the nonlinear equations

$$\begin{cases} \tau\, dX/dt = F(X,T) \\ \tau\, dT/dt = G(X,T) \end{cases} \qquad [2.51]$$

can we fully describe the unsteady behavior of the reactor and provide the response to large disturbances (Figure 2.8).

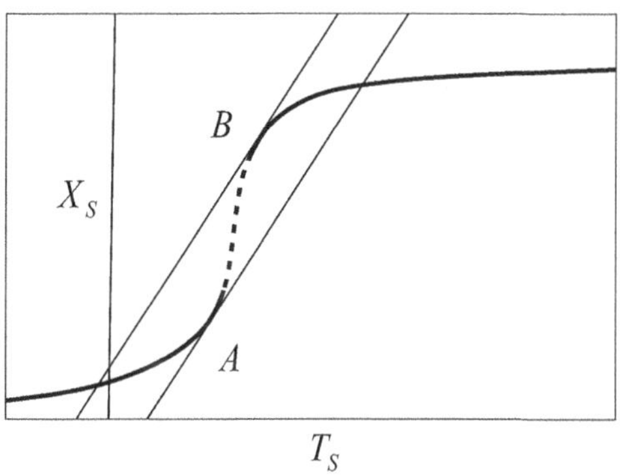

Figure 2.8. *Unstable steady solutions*

In the footsteps of Bilous and Amundson [BIL 55], we can take account of the evolution by giving a representation in the plane (T, X), divided into regions on the basis of the signs of $F(X,T)$ and $G(X,T)$.

Figure 2.9 is a qualitative illustration of what happens in the presence of a single steady solution. With three operating points, we obtain a separating line *xy* (Figure 2.10).

Figures 2.9 and 2.10 are obtained for a reversible reaction A $\rightleftarrows$ B, which is slightly different from the previous situation.

With the irreversible reaction $A \rightarrow B$, Aris and Amundson [ARI 58; ARI 65] numerically obtained the solutions in the presence of three steady points (Figure 2.11).

We note that if the trigger point is to the right of the separating line *xy*, we see stabilization of the operation on the steady point S_1. If we wish to obtain the stable steady point S_3, which is the only factor of interest, we will – as stated above – have to pre-heat the reactor, for instance.

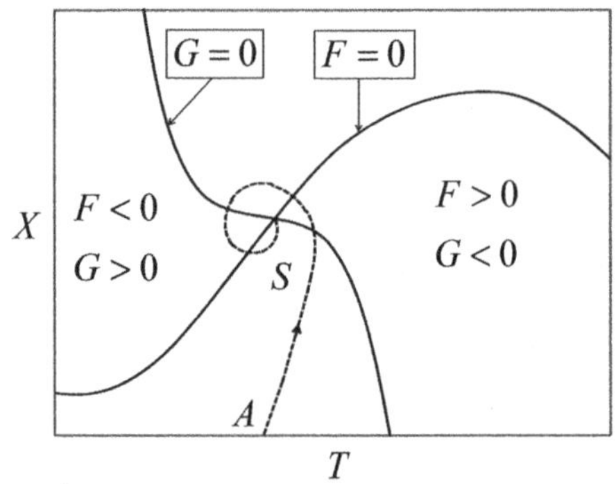

Figure 2.9. *Case of a single steady point. The path AS indicates oscillating stability of the steady point*

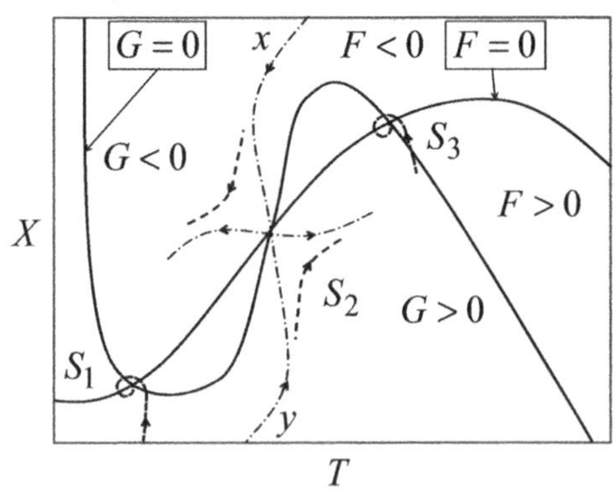

Figure 2.10. *Case of three steady points*

Two problems arise with this type of reactor:

– the time necessary to reach a stable steady state;

– the existence of unstable steady points.

These problems can be solved by control of the reactor, enabling us either to accelerate the progression toward the steady state or to stabilize a naturally unsteady operation.

We can, for instance, adjust the flowrate of the external cooling fluid, depending on the output temperature of the reactor. This enables us to affect the heat exchange coefficient K, which becomes $K + m$.

The conditions for stability of the linearized system therefore become

$$\begin{cases} 1+a_1+a_3+m-a_2a_4>0, \\ (1+a_1)(a_3+m)-a_2\,a_4>0 \end{cases} \qquad [2.52]$$

We note that if m takes a sufficiently high value, it will always be possible to stabilize the operating point.

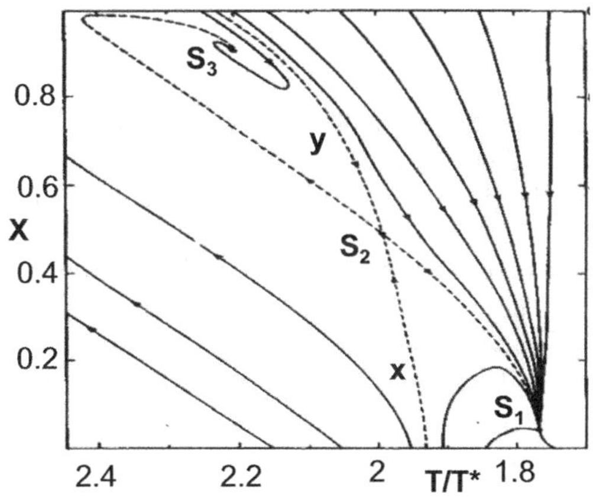

Figure 2.11. *Numerical solution for three steady states (redrawn from [ARI 58])*

However, in practical terms, it is difficult to make such an adjustment. We can also obtain stabilization, but only in certain situations, by using a signal based on the concentration rather than the temperature of the reactor. The heat flux with the exchanger then becomes a function $Q(T, X)$.

Linearization then gives us

$$\begin{cases} dx_1/d\theta = -(1+a_1)x_1 + a_1 a_2 x_2 \\ dx_2/d\theta = -(a_4+\mu)x_1 + (a_2 a_4 - a_3)x_2 \end{cases} \qquad [2.53]$$

where μ characterizes the influence of the concentration on Q, which is taken to be linear. The conditions of stability become

$$\begin{cases} 1 + a_1 + a_3 - a_2 a_4 > 0 \\ (1+a_1)a_3 - \mu a_1 a_2 > 0 \end{cases} \qquad [2.54]$$

We can see that the control only affects the second condition, i.e. it affects *P*. We can only stabilize points for which the first condition is satisfied.

Aris and Amundson (see [ARI 65] as well as [ARI 58; BIL 55]) carried out a numerical examination of the non-linearized system, with increasing values of *T* for the last reactor cited. They observed the stabilization of S_2 and, for intermediary values of μ, the apparition of boundary cycles (see figures in [PRU 10]).

2.3. Tubular reactor

A tubular reactor is a long cylindrical tube in which there is a flow with a chemical reaction. In a tubular reactor, the opening and exit are at the ends of the tube. The properties of the fluid flow vary in the axial and radial directions.

Homogeneous tubular reactors have a clearly defined residence time and good temperature control. They are used in particular for thermal cracking of hydrocarbons, the chloration of methane, propene and butadiene, the polymerization of ethylene into LLDPE – Linear Low Density PolyEthylene (exothermic reactions).

There are also heterogeneous, polyphasic, or in-fluidized-bed tubular reactors.

A great deal of attention has been paid to this type of reactor, used in hydrodesulfurization as mentioned above, as demonstrated by the theses [PEL 10; DOS 08] and the abundance of recent publications on the subject [AND 99; BAB 03; BRU 05; EXO 08; FRO 04; GIS 02; HAL 01; MAR 02; NOC 00; PLA 03; ROC 02; SON 03; TOP 96; WAU 94].

Many different simplifying hypotheses are possible for tubular reactors.

In the case of a plug flow reactor, the flow takes place in homogeneous planar layers, with no diffusive exchange with the upstream or downstream neighboring layers. However, there is an increase in the mass of the species R_j per unit of time because of one or more chemical reactions, and a global quantity of heat can be applied to the layer [PAR 69].

If we accept that diffusive exchanges take place, these may be radial or axial, or both in some cases [MCG 71].

Some of the output fluid can be recycled [PAR 69; BOR 05] by reinjecting it at the opening, possibly after passing it through a heat exchanger and mixing it with the input fluid.

There have been many articles published on the search for steady operating points and the study of their stability [MCG 71; BOS 02; KOR 03; BOY 05; DRA 08].

2.3.1. *Plug flow reactor*

A plug flow reactor (also called "piston reactor") is one of the ideal reactors mentioned in section 2.1.1.

We assume that the velocity of the flow is constant, and that there is only one chemical reaction

$$A \xrightarrow{k(T)} B \qquad [2.55]$$

The dimensionless equations are reduced to [PAR 69]:

$$\begin{cases} \dfrac{\partial C}{\partial t} + \dfrac{\partial C}{\partial x} = R(C,T) \\ \dfrac{\partial T}{\partial t} + \dfrac{\partial T}{\partial x} = -R(C,T) + K\left(T_p - T\right) \end{cases} \qquad [2.56]$$

with the reference values: $L_{ref} = L$ (length of the reactor), $C_{ref} = C_{A0}$ (concentration of reagents at opening), $T_{ref} = \Delta H\, C_{A0} / \rho C_p$. The second equation involves a linear heat exchange law where T_p is the temperature at the wall of the exchanger and K is the dimensionless exchange coefficient.

2.3.1.1. *Characteristics*

If we set $x-t=\xi, x+t=\eta$, so $x=(\xi+\eta)/2, t=(\xi-\eta)/2$, system [2.56] becomes

$$\begin{cases} \dfrac{\partial C}{\partial \eta}=\dfrac{R}{2} \\ \dfrac{\partial T}{\partial \eta}=\dfrac{-R+K\left(T_p-T\right)}{2} \end{cases} \qquad [2.57]$$

2.3.1.2. *Isothermic scenario*

With a 1st-order isothermic reaction

$$\frac{\partial C}{\partial t}+\frac{\partial C}{\partial x}=-k\,C \Rightarrow \frac{\partial C}{\partial \eta}=-\frac{k}{2}C \qquad [2.58]$$

We integrate equation [2.58], giving us

$$C(x,t)=A(\xi)e^{-\frac{k}{2}\eta}=A(x-t)e^{-\frac{k}{2}(x+t)} \qquad [2.59]$$

At $x=0$, $C(0,t)=C_0(t)=A(-t)e^{-\frac{k}{2}t}$, where $C_0(t)$ is known. It follows from this that $A(x-t)=C_0(t-x)e^{-kx}$. Thus, we obtain the reduced concentration as a function of the abscissa value and of time. At the reactor output, $x = 1$ and

$$C_e(t)=C(1,t)=C_0(t-1)\,e^{-k} \qquad [2.60]$$

2.3.1.3. *Non-isothermic scenario*

When $K=0$ (in the absence of an exchanger), we find $R=-B\,C\,e^{-T_a/T}$ and given [2.57]:

$$\begin{aligned} &C_e(t)+T_e(t)=C(1,t)+T(1,t)=C_0(t-1)+T_0(t-1) \\ &2\frac{\partial T}{\partial \eta}=B\left[C_0(-\xi)+T_0(-\xi)-T\right]e^{-T_a/T} \end{aligned} \qquad [2.61]$$

This last equation needs to be integrated with $\xi = const.$, so that

$$\frac{e^{T_a/T}\,dT}{C_0\left(-\xi\right)+T_0\left(-\xi\right)-T}=\frac{B}{2}d\eta \qquad [2.62]$$

The following tubular reactors are “less ideal” than the plug flow reactor. They are more or less faithful models of real reactors.

2.3.2. *Reactor with axial mixing*

2.3.2.1. *General points*

We suppose the tubular reactor with axial mixing to have a flat velocity profile, and not to exhibit any radial gradient of concentration or temperature. Equations [2.56] are replaced with the following [MCG 71]:

$$\begin{cases} \dfrac{\partial C}{\partial t}+\dfrac{\partial C}{\partial x}-Pe_C\dfrac{\partial^2 C}{\partial x^2}=R\left(C,T\right) \\ \dfrac{\partial T}{\partial t}+\dfrac{\partial T}{\partial x}-Pe_T\dfrac{\partial^2 T}{\partial x^2}=-R\left(C,T\right)+K\left(T_p-T\right) \end{cases} \qquad [2.63]$$

which involve the Peclet coefficients of mass transfer $Pe_C = vL/D$ and heat transfer $Pe_T = vL/\kappa$.

2.3.2.2. *Example of resolution with negligible barycentric velocity*

Below, we discuss a simplified case of a tubular reactor [CAL 77] with negligible barycentric velocity.[4] We consider a highly porous medium with a solid phase at rest, with density ρ_s, serving as a support for a second, condensed reactive phase with density ρ_R. This ensemble also includes the products of gaseous decomposition with partial densities ρ_j^g. Although the immobile condensed phases are the most dense, we can describe the medium using average quantities such as the total density and the barycentric velocity defined by $\vec{\mathbf{v}} = \sum_j Y_j^g \vec{\mathbf{v}}_j^g$. This latter is negligible because the mass fractions of the gaseous components are very small.

4 We wish to study the impact of the exothermic decomposition of a binder on the behavior of a piping heat protection layer.

The heterogeneous medium then behaves like a homogeneous medium, and the equations from system [2.63] are replaced with the following ones, which are written thus, in their dimensionless form:

$$\begin{cases} \dfrac{\partial C}{\partial t} - D\dfrac{\partial^2 C}{\partial x^2} = -R(C,T) \\ \dfrac{\partial T}{\partial t} - \kappa\dfrac{\partial^2 T}{\partial x^2} = -\dfrac{\Delta h}{\rho c}R(C,T) \end{cases} \qquad [2.64]$$

On the right-hand side of the second equation, we neglect the product of the heat flux by diffusion flux (as done in Chapter 2 of [PRU 12], with the Shvab-Zeldovich approximation).

The chemical reaction, taken to be of order α, is expressed symbolically by

$$\mathcal{R} \xrightarrow{k(T)} \sum_j \nu_i^g \, \mathcal{E}_j^g$$

with the left-hand side denoting the reagents. Its rate is written as

$$R(C,T) = BT^s C^\alpha e^{E_a/RT} \qquad [2.65]$$

The boundary conditions are as follows:

$$\begin{cases} x = 0: \dfrac{\partial C}{\partial x} = 0,\ T = T_o(t) \\ x = l: \dfrac{\partial C}{\partial x} = 0,\ \lambda\dfrac{\partial T}{\partial x} = b(T - T_0) \end{cases} \qquad [2.66]$$

The temperature $T_o(t)$ is a parietal temperature and the temperature T_0 of the surrounding environment is supposed to be constant.

The dimensionless equations are placed in the following form:

$$\begin{cases} Le\dfrac{\partial C}{\partial t} - \dfrac{\partial^2 C}{\partial x^2} = -\varphi^2 C e^{\gamma\left(1-\frac{1}{T}\right)} \\ \dfrac{\partial T}{\partial t} - \kappa\dfrac{\partial^2 T}{\partial x^2} = -\varphi^2 \beta C e^{\gamma\left(1-\frac{1}{T}\right)} \end{cases} \qquad [2.67]$$

and reveal the dimensionless numbers (Lewis numbers): $\varphi^2 = \frac{Bl^2}{D} e^{-E_a/RT_0}$, $\beta = \frac{-\Delta h\, DC_0}{\lambda T_0}, \gamma = \frac{E_a}{RT_0}$. The reference values are

$$x_{ref} = l,\; t_{ref} = \kappa l^2,\; C_{ref} = C_0,\; T_{ref} = T_0$$

The boundary conditions become

$$\begin{cases} x = 0: \dfrac{\partial C}{\partial x} = 0,\; T = T_e(t) \\ x = 1: \dfrac{\partial C}{\partial x} = 0,\; \lambda \dfrac{\partial T}{\partial x} = Bi(T-1) \end{cases} \quad [2.68]$$

with the Biot number $Bi = b/\lambda$.

The system of equations thus obtained is solved by way of the finite difference method (Crank-Nicolson scheme), taking the following to be the initial conditions:

$$T(0,0) = T_e \text{ and } C(0,0) = 1;\;\; T(x,0) = 1 \text{ and } C(x,0) = 1 \text{ for } x > 0$$

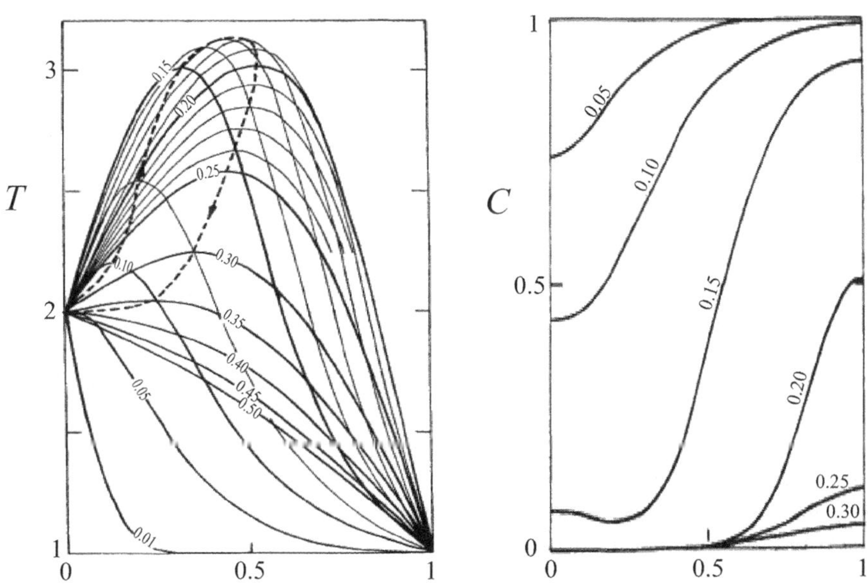

Figure 2.12. *Fields of temperature $T(x,t)$ and concentration $C(x,t)$, graduated curves in reduced time, obtained for $Bi = 1000, \gamma = 15, \varphi = 0.01, \beta = 2$ ([CAL 77])*

The results in Figure 2.12 show that, for given values of the parameters, there is progression of an *extremum* of temperature, which increases and then falls off as the reactive species decomposes.

2.3.3. *Reactor with radial mixing*

This time, we assume that the tubular reactor has a planar velocity profile and does not exhibit an axial gradient for concentration or temperature. The equations [2.56] are replaced with the following [MCG 71]:

$$\begin{cases} \dfrac{\partial C}{\partial t} + \dfrac{\partial C}{\partial x} - Pe_C \dfrac{1}{r}\dfrac{\partial}{\partial r}\left[r \dfrac{\partial C}{\partial r} \right] = R(C,T) \\ \dfrac{\partial T}{\partial t} + \dfrac{\partial T}{\partial x} - Pe_T \dfrac{1}{r}\dfrac{\partial}{\partial r}\left[r \dfrac{\partial T}{\partial r} \right] = -R(C,T) \end{cases} \qquad [2.69]$$

This time, the term of heat exchange through the wall does not appear on the right-hand side of the second equation, but is present in the boundary conditions:

$$\begin{aligned} & x = 0 : C(0,r) = 1,\ T(0,r) = T_F \\ & r = 0 : \left.\frac{\partial C}{\partial r}\right|_0 = \left.\frac{\partial T}{\partial r}\right|_0 = 0 \\ & r = 1 : \left.\frac{\partial C}{\partial r}\right|_1, \ \frac{1}{Pe_T}\left.\frac{\partial T}{\partial r}\right|_1 = K\left[T_p - T(1)\right] \end{aligned} \qquad [2.70]$$

These systems of equations are solved numerically.

It is not always easy to characterize a real reactor in industrial installations, because we often have access only to the opening and exit sections.

Generally, we are interested in the quality of the mixture (section A3.1.4), because efficient micromixing ensures intimate contact on the molecular scale between the reactants, facilitating chemical reactions. We are also interested in the residence time of the components in the reactor. These notions are related to that of the scale of turbulence (Appendix 3 and Chapter 3), with small scales being responsible for micromixing whereas large scales tend more to affect the residence times.

In the next section, we shall see that it is possible to determine certain characteristics by analyzing the output signal from a reactor fed with neutral fluid, i.e. with no chemical reactions, into which a tracker is injected. For other characteristics, we need to simulate the turbulence.

2.4. Residence time distribution

2.4.1. *Balance equations*

In the case of real reactors, we can use the probabilistic balance equation [2.9]. For a reactor with constant volume υ, and opening and exit volumetric flowrates $\dot{q}_o$ and $\dot{q}_e$, this equation becomes

$$\frac{\partial(\overline{\psi})}{\partial t}+\frac{\dot{q}_e\overline{\psi}_e-\dot{q}_o\overline{\psi}_o}{\upsilon}+\sum_i\frac{\partial(\overline{\psi}w_i)}{\partial\zeta_i}=\overline{G} \quad [2.71]$$

where ψ is the distribution function of the phase space:

$$\psi=\psi\left(\vec{\mathbf{x}},\vec{\zeta},t\right) \quad [2.72]$$

The averages $\overline{\psi},\overline{G},\overline{\psi}_o$ and $\overline{\psi}_e$ are defined by

$$\begin{aligned}&\overline{\psi}=\frac{1}{\upsilon}\int_\upsilon\psi\,d\mathcal{V},\ \overline{G}=\frac{1}{\upsilon}\int_\upsilon G\,d\mathcal{V},\\&\overline{\psi}_o=\frac{1}{\dot{q}_o}\int_{A_o}\psi\vec{\mathbf{v}}\cdot d\vec{\mathbf{A}},\ \overline{\psi}_e=\frac{1}{\dot{q}_e}\int_{A_e}\psi\vec{\mathbf{v}}\cdot d\vec{\mathbf{A}}\end{aligned} \quad [2.73]$$

A_o and A_e being the opening and exit sections of the reactor.

Many models have been elaborated on the basis of a choice of parameters ζ_j and the giving of the corresponding $\dot{w}_j$ values and the average production $\overline{G}$.

At an experimental level, we can make use of the notions of age and residence time of the molecules in the reactor. We neglect the influence of the chemical reactions on the flow and use a neutral fluid to characterize that flow. Thereby we define the age distribution α such that $I(\alpha)d\alpha$ is the fraction of the molecules in the reactor whose age is between α and $\alpha+d\alpha$. The residence time distribution

t_e, or $E(t_e)$, is such that $E(t_e)dt_e$ represents the output flowrate fraction of the molecules having spent a time-period between t_e and $t_e + dt_e$ in the reactor.

The residence time distribution can be determined experimentally by using a tracer whose concentration or flowrate is measured in the exit area A_e. If an impulse of the tracer (in a Dirac peak) is introduced at time $t = 0$ in the opening area, $C(x_s, t_s)$ being the concentration of the tracer in the output area, we would have

$$E(t_e) = \frac{\int_{A_e} C(x_e, t_e)\vec{\mathbf{v}}(x_e, t_e)\cdot d\vec{\mathbf{A}}}{\int_{t_e=0}^{\infty}\int_{A_e} C(x_e, t_e)\vec{\mathbf{v}}(x_e, t_e)\cdot d\vec{\mathbf{A}}} \qquad [2.74]$$

The passage time τ is equal to the average residence time:

$$\tau = \overline{t_e} = \int_0^{\infty} t_e E(t_e)dt_e \qquad [2.75]$$

Let us now give three examples drawn from case studies with clearly determined reactors.

2.4.2. *Perfectly stirred homogeneous reactors in a steady regime*

In the absence of a reaction, we have

$$dC/dt = (C_o - C)/\tau \qquad [2.76]$$

for $t > 0$, with τ being the passage time equal to $\mathcal{V}/\dot{q}$. Given that $C_o = C_o\,\delta(t)$, the solution becomes

$$C = C_0 e^{-t/\tau}, t > 0 \qquad [2.77]$$

Hence, if v represents the velocity of the flow, we have

$$\int_{A_s} C v dA = C_0 e^{-t/\tau}\int_{A_s} v dA = C_0 \dot{q} e^{-t/\tau} \qquad [2.78]$$

so that

$$E(t_e) = \frac{e^{-t_e/\tau}}{\int_0^\infty e^{-t_e/\tau}\,dt_e} = \frac{1}{\tau} e^{-t_e/\tau} \qquad [2.79]$$

2.4.3. *Plug flow reactors*

A plug flow reactor is a tubular reactor in which the flow takes place in flat layers in which the parameters are uniform. A pulsed injection of dye at the opening, in the absence of diffusion, will yield the same signal at exit, after a passage time *t*, so that

$$E(t_e) = \delta(t_e - \tau) \qquad [2.80]$$

2.4.4. *Poiseuille flow*

The velocity profile is given by:

$$u(r) = \frac{\Delta p}{4\mu\,\Delta l}\left(R^2 - r^2\right),\ R = \frac{D}{2} \qquad [2.81]$$

The residence time is therefore a function of *r*:

$$t_e = \Delta l / u(r) = \Delta l / a\left(R^2 - r^2\right),\ a = \Delta p / 4\mu\,\Delta l \qquad [2.82]$$

The flowrate between t_e and $t_e + dt_e$ is

$$u(r)\, 2\pi r\, dr = 2\pi\, a\, r\left(R^2 - r^2\right) dr \qquad [2.83]$$

and, by integration between 0 and *R*, the total flowrate becomes

$$2\pi\, a R^4 / 4 \qquad [2.84]$$

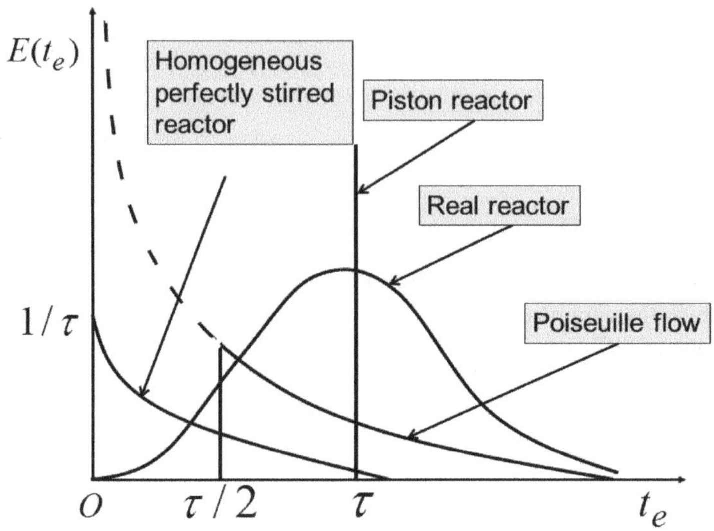

Figure 2.13. *Residence time distributions for various chemical reactors*

Thus

$$E(t_e)dt_e = 4u(r)r\,dr/aR^4 = 4(R^2 - r^2)r\,dr/R^4 \qquad [2.85]$$

Because

$$dt_e = \frac{\Delta l}{a}\frac{2r\,dr}{(R^2 - r^2)^2} \qquad [2.86]$$

we find

$$\begin{cases} E(t_e) = \tau^2/2t_e^3,\ \tau/2 \le t_e < \infty, \\ E(t_e) = 0,\ 0 \le t_e < \tau/2 \end{cases} \qquad [2.87]$$

where

$$\tau = \overline{t_e} = 2\Delta l/aR^2 \qquad [2.88]$$

The solutions for these three examples are represented in Figure 2.13, along with the shape of the residence time distribution for a particular real reactor. The differences between the different types of ideal reactors are manifest and constitute indications to characterize a given reactor.

2.4.5. *Real reactors*

The age distribution and residence time distribution can be linked by the population balance equation. The age α then characterizes a particle of fluid, such that $\zeta = \alpha$ and $w = d\alpha/dt = 1$ in steady flow.

We then have

$$\frac{\dot{q}_e \bar{\psi}_e}{\mathcal{V}} + \frac{\partial \bar{\psi}}{\partial \alpha} = 0, \; \alpha > 0 \qquad [2.89]$$

However, the age distribution is equal to $\bar{\psi}/N$ where $N = \int_0^\infty \bar{\psi}\, d\alpha$, and the residence time distribution is given by $\bar{\psi}_e/N$.

Thus, if we set $\tau = \mathcal{V}/\dot{q}_e$, for an incompressible fluid, we obtain

$$dI(\alpha)/d\alpha + E(\alpha)/\tau = 0 \qquad [2.90]$$

For instance, with a perfect mixture, we find

$$I(\alpha) = E(\alpha) = e^{-\alpha/\tau}/\tau \qquad [2.91]$$

In general

$$\tau I(t) = 1 - \int_0^t E(t_e)\, dt_e \qquad [2.92]$$

It is easy to interpret this relation if we imagine an injection of dye into the incompressible fluid, in unitary stages. By integration, we find

$$\mathcal{V} \int_0^t I(\alpha)\, d\alpha = \int_0^t \dot{q}\, dt' \int_{t'}^\infty E(t_e)\, dt_e \qquad [2.93]$$

where $\int_{t'}^{\infty} E(t_e)\,dt_e$ represents the fraction of the output flowrate that is older than t', and therefore not colored. Hence, this is a balance at time $t > 0$: the quantity of colored fluid existing in the reactor at time t is equal to the quantity of non-colored fluid ejected after the injection of the dye has commenced ($t = 0$).

The residence time distribution characterizes the macromixing, i.e. the displacement of particles of fluid in relation to one another.

The micromixing, for its part, relates to the interactions and exchanges of the content of the aggregates[5] with their environment. The micromixing depends heavily on the state of small-scale turbulence. Therefore, it is complicated to study. Specialists in chemical engineering have developed models with a different approach to that of fluid mechanics specialists.

In their IEM (interaction by exchange with the mean) model for instance, the concentration C of a chemical aggregate satisfies an equation:

$$dC/d\alpha = h\left(C - \bar{C}\right) + \dot{C}(C) \qquad [2.94]$$

where h is a frequency of exchange and $\bar{C}$ is the average concentration:

$$\bar{C} = \int_0^{\infty} I(\alpha)C(\alpha)\,d\alpha \qquad [2.95]$$

The average production rate $\bar{\dot{C}}$ is

$$\bar{\dot{C}} = \int_0^{\infty} I(\alpha)\dot{C}\left[C(\alpha)\right]d\alpha \qquad [2.96]$$

This gives us an integrodifferential equation:

$$dC/d\alpha = h\left(C - \int_0^{\alpha} I(\alpha)C(\alpha)\,d\alpha\right) + \dot{C}(C) \qquad [2.97]$$

the solving of which requires knowledge of the initial probability density of concentration.

5 All real fluids are partially composed of molecular aggregates, formed by the shearing effect due to the stirring and mechanical friction. These aggregates exchange matter with one another by diffusion.

Chapter 3

Laminar and Turbulent Flames

Combustion is the best-known and most problematic application for the general equations of reactive fluids (see [PRU 12], section 2.3). We shall first discuss the *laminar regime* and then the *turbulent regime* and, in each case, we shall examine *premixed combustion* and then *non-premixed combustion* in turn.

Here, we shall not deal with the questions of ignition and self-ignition to their full extent, but we do wish to stress their importance, because they are of particular interest with regard to the operation of engines.[1] We shall touch on the issue of *triple flames*, which is applicable to this discussion, in section 3.3.6.

3.1. Laminar premixed combustion

To begin with, by examining the one-dimensional case of flow in a tube with constant cross-section, we shall show that there are two steady regimes for premixed combustion (Rankine-Hugoniot theory):

---

1 Indeed, anyone who speaks of combustion also speaks of *ignition*; this may be due to the fact of a hot point that triggers the exothermic chemical reaction, the heat produced then sustaining the propagation of a deflagration wave (in a so-called internal combustion engine using gasoline as a fuel) or a detonation wave in a premix [JAU 71], or the development of a diffusion flame on contact between a separate fuel and oxidizer (a candle flame). However, the ignition of a mixture of fuel and oxidizer may also be spontaneous at sufficiently high temperature – this is *self-ignition* (explosion, or a Diesel engine). Many different modes of progression are envisageable [ZEL 80]. In principle, if the temperature and concentration conditions are homogeneous, the fluid will ignite almost simultaneously at all points; yet in reality there are almost always hot points [GU 03].

– deflagration, which occurs at quasi-uniform pressure and which corresponds to a plane premixed flame, propagating at a slow speed (a few centimeters to a few meters per second);

– detonation, which is a genuine combination between shock and combustion, which causes a pressure jump when it occurs and propagates at supersonic speed.

Then, we shall look at the inside of a deflagration wave, which is the locus of an exothermic chemical reaction with high activation energy, and of phenomena of diffusion of mass and heat, whilst the viscous effects are ignored.

3.1.1. *Rankine-Hugoniot theory*

In a sufficiently long straight tube filled with a mixture of fuel and oxidizer, if we ignite the mixture at one end of the tube, a deflagration flame propagates at a constant speed for a given length of time. When that flame has covered a distance of around ten times the diameter of the tube, the flame gathers speed and is transformed, after a transitory period, into a detonation wave with constant velocity. Here, we shall only study plane waves for which the motion of the fluid can be considered to be one-dimensional and steady in a framework relating to the wave. Thus, we shall not take account of the transitory period in our study.

In a system of axes linked to the wave, the phenomenon can be represented diagrammatically as follows (Figure 3.1a):

– in front of the wave ($x < 0$), the mixture of fresh gases moves at a constant velocity v_1;

– at the wave front ($x = 0$) the fluid experiences a discontinuity;

– behind the wave ($x > 0$) the mixture of burned gases moves uniformly at velocity v_2.

Conversely, if we consider a system of axes in relation to which the fresh gases are immobile, the wave would work its way back up through the fresh mixture (Figure 3.1b).

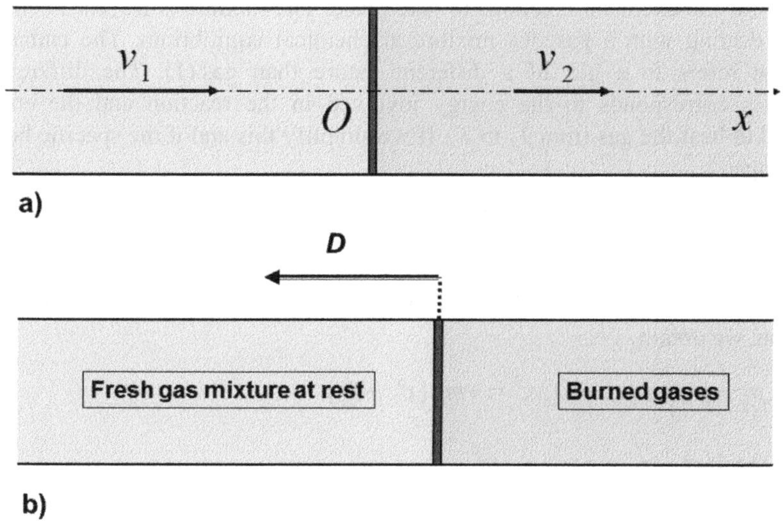

Figure 3.1. *System of axes: a) relative to the wave, b) the wave progresses through the fresh mixture*

The equations at the discontinuity are (as explained in [PRU 12]), successively for the mass, momentum and energy:

$$\rho_1 v_1 = \rho_2 v_2 = \dot{m} \qquad [3.1]$$

$$p_1 + \rho_1 v_1^2 = p_2 + \rho_2 v_2^2 \qquad [3.2]$$

$$h_1 + v_1^2/2 = h_2 + v_2^2/2 \qquad [3.3]$$

To these motion equations, we have to add the equations of state of the gaseous mixture. There are two of these equations in the case of perfect gases in an ideal mixture[2]; we have

$$\begin{cases} p_1/\rho_1 = n_1 R T_1,\ p_2/\rho_2 = n_2 R T_2, \\ h_1 = h_1\left(T_1, p, Y_{j1}\right),\ h_2 = h_2\left(T_2, p, Y_{j2}\right) \end{cases} \qquad [3.4]$$

2 On this point, see section 2.2.2 of [PRU 12], and particularly footnote 7. A mixture of perfect gases is generally ideal.

In state (1), the concentrations are those of the fresh gases, and the temperature is too low for the chemical reactions to take place. The mixture is frozen. In state (2), we are dealing with a gaseous mixture at chemical equilibrium. The enthalpy h_2 therefore refers to a gas of a different nature than gas (1). The difference in enthalpies corresponds to the energy involved in the reaction and the enthalpy required to heat the gas from T_1 to T_2. If we simplify this and if the specific heat has not changed[3]:

$$h_2 - h_1 = -\Delta h + c_p \left(T_2 - T_1\right) \qquad [3.5]$$

By eliminating v_1 and v_2 between the continuity equation and the momentum equation, we obtain

$$p_2 - p_1 = \rho_1 v_1^2 - \rho_2 v_2^2 = -\dot{m}^2 \left(\vartheta_2 - \vartheta_1\right) \qquad [3.6]$$

where $\vartheta = 1/\rho$, or

$$\left(p_2 - p_1\right)/\left(\vartheta_2 - \vartheta_1\right) = -\dot{m}^2 \qquad [3.7]$$

By doing the same thing to the energy equation, we obtain

$$h_2 - h_1 = -\left(\dot{m}^2/2\right)\left(\vartheta_2^2 - \vartheta_1^2\right) \qquad [3.8]$$

and, by replacing $\dot{m}^2$ with its value

$$h_2 - h_1 = -(1/2)\left(\vartheta_2 + \vartheta_1\right)\left(p_2 - p_1\right) \qquad [3.9]$$

This is the *Hugoniot adiabatic curve* in the plane (ϑ, p). Equations [3.7] and [3.9] are called the "*Rankine–Hugoniot relations*".

In the system of axes $\left(\vartheta_2, p_2\right)$, equation [3.7] gives us a straight line (the so-called "*Hugoniot straight line*") with slope $\left(-\dot{m}^2\right)$. The quantities ϑ_1, p_1 are data, and ϑ_2, p_2 are the unknowns in the problem.

---

3 NOTE.– At the wave front of a shock without combustion, we have only $h_2 - h_1 = c_p\left(T_2 - T_1\right)$.

We saw earlier that $h_2 - h_1$ could be expressed as a function of the thermodynamic values (e.g. the temperature T); if the variables chosen are p and ϑ, equation [3.9] is represented by a curve that does not pass through the point (ϑ_1, p_1); indeed, if for instance we take $T_2 = T_1$, we do not obtain $h_2 = h_1$, unlike what happens with a shockwave.

The intersection of the adiabatic curve defined by equation [3.9] and the line with slope $-\dot{m}^2$ gives the solution.

We are going to solve this problem in the simplified case mentioned above. In addition, we shall suppose that the molar mass does not change, so that $T_2/T_1 = p_2\,\vartheta_2/p_1\,\vartheta_1$. We therefore set $\vartheta_2/\vartheta_1 = \vartheta, p_2/p_1 = p, \tau = \Delta h/c_p T_1$.

In light of equations [3.7] and [3.9], we obtain

$$\begin{cases} \dfrac{p-1}{\vartheta-1} = -\dot{m}^2 \dfrac{\vartheta_1}{p_1}, \\ p\,\vartheta - 1 - \tau = \dfrac{p_1\,\vartheta_1}{2c_p\,T_1}(\vartheta+1)(p-1) \end{cases} \qquad [3.10]$$

In view of the expression of the Mach number $M_1 = v_1/c_1 = v_1/\sqrt{\gamma\,p_1\,\vartheta_1}$ and the laws of state of gases, we have

$$\begin{cases} \dfrac{p-1}{\vartheta-1} = -\gamma M_1^2, \\ p = \dfrac{(1-\gamma)\vartheta + 1 + \gamma + 2\gamma\tau}{(1+\gamma)\vartheta + 1 - \gamma} \end{cases} \qquad [3.11]$$

System [3.11] gives us the *Hugoniot straight line for the mass flow rate* and *Hugoniot adiabatic curve* shown in Figure 3.2.

The prohibited zone corresponds to $M_1^2 < 0$. The part $\vartheta < 1$ of this curve corresponds to an increase in pressure – often a significant jump. This is the detonation domain. When $\tau = 0$, i.e. in the absence of combustion, we are simply dealing with a shockwave. We can demonstrate in this case, by applying the second law of thermodynamics, that the shockwave is necessarily a compression wave. Hence, there cannot be a discontinuity wave, in a simple fluid, with increasing specific volume and decreasing pressure. The same is not true when τ is positive.

The domain corresponding to $p < 1$ is the deflagration domain. Generally, the variation in pressure is so slight as to be negligible. Hence, we suppose the pressure to be constant when studying plane deflagrations.

Thus, we can show two distinct domains – the detonation and the deflagration domain. We shall see that the discontinuity condition shown in Table 2.2 in Chapter 2 of [PRU 12], applied to entropy, which here gives us

$$\left[\vec{\jmath}_S + \rho s\left(\vec{\mathbf{v}} - \vec{\mathbf{W}}\right)\right]_-^+ \cdot \vec{\mathbf{N}} = \dot{W}_{aS} \geq 0 \qquad [3.12]$$

leads to the further limitation of these domains. Let us first define the two specific points where the straight line issuing from point (1, 1) is tangent to the adiabat. Firstly, in the vicinity of these points, we can differentiate equations [3.11], which gives us the following:

$$dp = -\gamma M_1^2\, d\vartheta, \left[(1+\gamma)\vartheta + 1 - \gamma\right]dp + p(1+\gamma)d\vartheta = (1-\gamma)d\vartheta.$$

By eliminating dp we get $-\gamma M_1^2\left[(1+\gamma)\vartheta + 1 - \gamma\right] + p(1+\gamma) - (1-\gamma) = 0$, which, if we replace M_1^2 with its value in p and ϑ, becomes $p\left[(1+\gamma)\vartheta - \gamma\right] - \vartheta = 0$. Finally, by eliminating p, we find equations which satisfy the abscissa values for these so-called Chapman–Jouguet points:

$$\vartheta^2 - 2(1+\tau)\vartheta + 1 + 2\gamma\tau/(1+\gamma) = 0 \qquad [3.13]$$

Let us now demonstrate that these *Chapman–Jouguet points* correspond to the *extrema* of entropy production. The entropy s_2 depends on the intersection of the straight line originating at (1,1) with the adiabat, and therefore on the slope of that line which, as we have seen, is proportional to the square of the upstream Mach M_1^2.

We shall now calculate the *extrema* of $(s_2 - s_1)$, i.e. the *extrema* of entropy production across the wave. We have

$$\dot{W}_{aS}/\dot{m} = s_2 - s_1 = c_v \ln(p_2/p_1) - c_p \ln(\rho_2/\rho_1) = c_v \ln p + c_p \ln \vartheta \qquad [3.14]$$

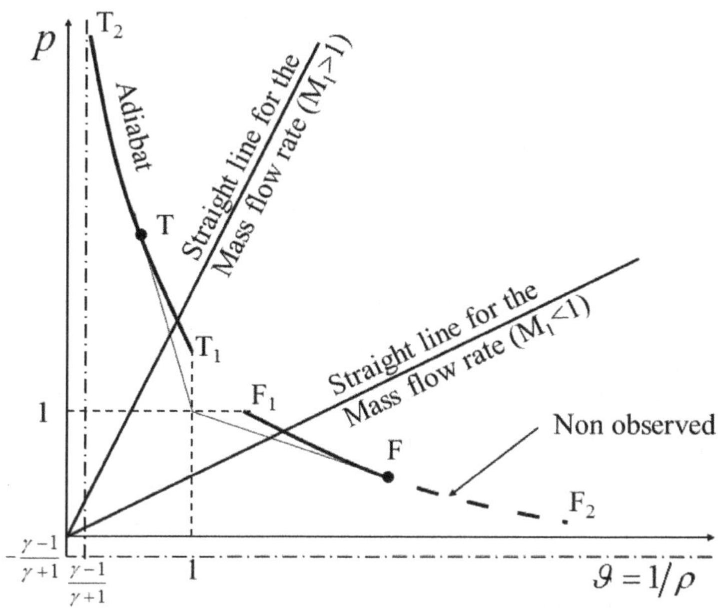

Figure 3.2. *Flowrate line and Hugoniot adiabat*

The *extremum* corresponds to $d(s_2 - s_1)/c_v = d\ln p + \gamma d\ln\vartheta = 0$. If we replace p with its value, we obtain

$$\frac{1}{c_v}\frac{d(s_2 - s_1)}{d\vartheta} = \frac{1-\gamma}{(1-\gamma)\vartheta + 1 + \gamma + 2\gamma\tau} - \frac{1+\gamma}{(1+\gamma)\vartheta + 1 - \gamma} + \frac{\gamma}{\vartheta}$$

or indeed equation [3.13].

We note that the *extrema* correspond to the Chapman–Jouguet points. In particular, detonation corresponds to the minimum entropy production at the Chapman–Jouguet point.

We can say that the detonation wave thus obtained is stable. However, the stability of the detonation wave generally depends on other factors such as the characteristics of the walls, and an in-depth study is needed to determine the stable zones. Such a study will not be performed here.

Looking at $(s_2 - s_1)/c_v$ as a function of ϑ, we find the shape of the curve in Figure 3.3. When $\tau = 0$, we again find the case of the shockwave characterized by a point of inflection in the horizontal tangent at $\vartheta = 1$.

The T_1 T_2 branch of the entropy production curve corresponds to detonation waves, and the F_1 F_2 branch to deflagration waves. The Chapman–Jouguet points are at T and F. Only point T has a particular meaning, corresponding to Chapman–Jouguet detonation.

The celerity of a detonation wave D is equal to the relative velocity of the fresh gases v_1. In particular, we can calculate that celerity at point T, where we have $\vartheta = 1 + \tau - \sqrt{\tau^2 + 2\tau/(\gamma+1)}$. We replace ϑ with this value in the expression of p and use the equation $M_1^{\ 2} = (p-1)/\gamma(1-\vartheta)$ to calculate M_1, from which we deduce $v_1 = D$ because $v_1 = M_1 c_1$, where c_1 is the speed of sound in the fresh gases, supposed to be known if the gases are at rest ($u_1 = 0 = v_1 - D$).

The downstream Mach number $M_2 = v_2/c_2$ in relation to the wave is such that:

$$M_2^{\ 2} = \frac{v_2^{\ 2}}{c_2^{\ 2}} = \frac{\rho_1^{\ 2} v_1^{\ 2}/c_1^{\ 2}}{\rho_2^{\ 2} c_2^{\ 2}/c_1^{\ 2}} = \frac{\rho_1^{\ 2} M_1^{\ 2}}{\rho_2^{\ 2}\, p_2\rho_1/p_1\rho_2} = \frac{\vartheta}{p} M_1^{\ 2} \qquad [3.15]$$

We can easily deduce the velocity v_2.

At the Chapman–Jouguet point T, we obtain $M_{2T} = 1$. In order to demonstrate this, we need only express $M_1^{\ 2}$ as a function of ϑ and p as before, and use the relation independent of τ established above, which is valid at the Chapman–Jouguet points:

$$p\left[(1+\gamma)\vartheta - \gamma\right] - \vartheta = 0$$

The fact that M_2 is equal to 1 in this case lends stability to the detonation wave. Indeed, the sound waves cannot move back up the flow to reach the detonation wave, and any slowing of this wave will be the source of an unstable phenomenon.

The zones of validity shown in Figures 3.2 and 3.3 (solid lines) illustrate a few elements deduced from theory and from experience. Note that, while the point T defining the Chapman–Jouguet detonation is stable (entropy production minimum) and does indeed correspond to the detonations observed, the same is not true of the point F. We know only that the deflagration wave will lie between the points F_1 and F on the adiabat. The Rankine–Hugoniot theory does not enable us to calculate the velocity of that wave.[4]

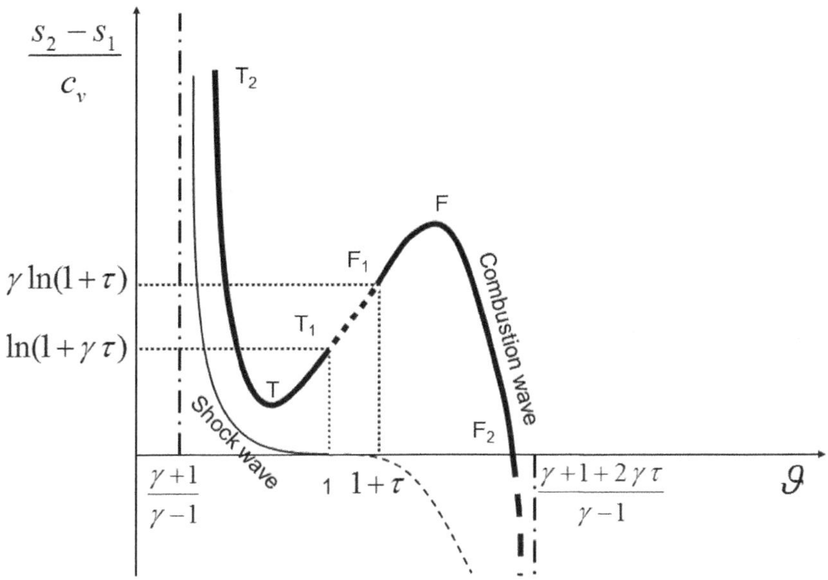

Figure 3.3. *Rate of entropy production, at the front of the combustion wave, along the Hugoniot adiabat. The shockwave is represented by the finer line*

3.1.2. *Velocity and structure of the plane adiabatic laminar and steady premixed flame*

Consider the plane deflagration wave defined in section 3.1.1, not as a simple discontinuity but rather as a thin zone where all of the dissipative phenomena take place: diffusion of chemical species, heat conduction and chemical reactions (Figure 3.4). This thin flame corresponds to the case of a chemical reaction with

4 Hence, in what follows, we shall limit ourselves to studying the structure of the deflagration wave, which is crucial for determining the combustion rate. The structure of the plane detonation wave, although less crucial, is also important, and has also been presented in [PRU 88; PRU 10]. It should also be noted that these detonation waves often have a three-dimensional structure.

high activation energy. Here, for simplicity's sake, we shall suppose that a single reaction summarizes the whole of the chemical process.

We shall now touch on this problem in the context of the Shvab-Zeldovich approximation presented in Chapter 2 of [PRU 12], and take a referential framework relating to the wave. All the parameters are dependent on a single variable x.

In a manner of speaking, this is a tubular reactor fed with fresh gas so that the flame is at rest, or indeed with a fresh mixture in which a flame propagates away from one open end where ignition took place.

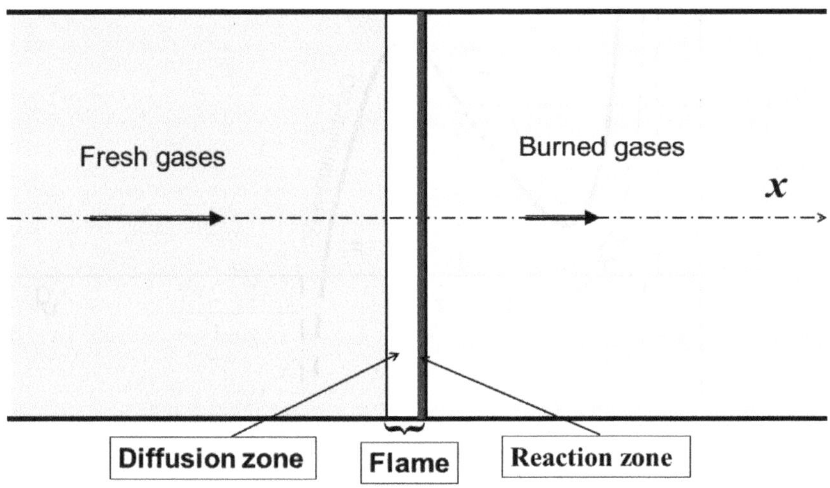

Figure 3.4. *Structure of the thin premixed flame*

Let us use the notation $\rho v = \dot{m}$ for the unitary mass flowrate and $\lambda / c_{pf} = \rho D = g$ for the common value of the diffusive transfer coefficients. We introduce the position variable η such that $d\eta / \eta = \dot{m}\, dx / g$.

We shall envisage the 2nd-order chemical reaction $A + B \to P$ and accept, for simplicity's sake, that before the reaction, one mole of A is combined with one mole of B. Using the reduced variables (β_j, β_T) defined in section 2.5.2 [PRU 12], we then have in $\eta = 0\,(x = -\infty)$:

$$\beta_A^0 = \beta_B^0 = -1 \qquad [3.16]$$

The equations are then as follows:

$$\begin{cases} \eta^2 \, d^2\beta_j / d\eta^2 = -g\,\dot{\zeta}/\dot{m}^2 \\ \eta^2 \, d^2\beta_T / d\eta^2 = -g\,\dot{\zeta}/\dot{m}^2 \\ \dot{\zeta} = B\,\rho^2 T^s \exp(-T_a/T)\,\beta_A\,\beta_B \end{cases} \quad [3.17]$$

We eliminate $\dot{\zeta}$ between the equations in (β_A, β_B), and obtain a linear relation giving $(\beta_A - \beta_B)$ as a function of η. If we accept that the solution is bounded in $\eta = +\infty \; (x = +\infty)$, this gives us

$$\beta_A = \beta_B = -\beta \quad [3.18]$$

We now eliminate $\dot{\zeta}$ between the equations in (β_A, β_T), and find $d^2(\beta + \beta_T)/d\eta^2 = 0$, which gives us the linear relation $\beta + \beta_T = a\eta + b$. The coefficients a and b in this relation are determined using the boundary conditions. For $\eta = 0, \beta = 1, \; \beta_T = 0 \Rightarrow b = 1$. For a large value of η, the reactions are finished and the variation of β_T can stem only from the exchanges with the external environment, and a is, in a manner of speaking, the adiabaticity of the reactor:

$$\eta = \infty, \quad \beta = 0, \quad d\beta/d\eta = 0 \Rightarrow a = d\beta_T/d\eta$$

In the *case of an adiabatic reactor* $a = 0$ and therefore we find:

$$\beta + \beta_T = 1 \quad [3.19]$$

We still need to solve a differential equation:

$$\eta^2 \frac{d^2\beta_T}{d\eta^2} = -\frac{g}{\dot{m}^2} B\,\rho^2 T^s e^{-T_a/T} (1 - \beta_T)^2 \quad [3.20]$$

Let $T_{ad} = T_0 + \sum_j \nu_j \mathcal{M}_j \left(q_f^0\right)_j / c_p$ represent the temperature at the end of adiabatic combustion; we have approximately $\beta_T = (T - T_0)/(T_{ad} - T_0)$. If we take a perfect gas law of state, $\rho = \bar{\mathcal{M}}\, p/RT$, where $\bar{\mathcal{M}}$ is the average molar mass, we obtain

$$\eta^2 \frac{d^2\beta_T}{d\eta^2} = -\Lambda\,\omega(\beta_T) \quad [3.21]$$

with

$$\begin{cases} \Lambda = g\, B\, \overline{\mathcal{M}}^2\, p^2\, T^s / \dot{m}^2\, R^2 \cong const. \\ \omega(\beta_T) = \left[\dfrac{1-\beta_T}{T_0 + (T_{ad} - T_0)\beta_T} \right]^2 \exp\left[-\dfrac{T_a}{T_0 + (T_{ad} - T_0)\beta_T} \right] \end{cases} \qquad [3.22]$$

Integration of the differential equation [3.21] presents a number of difficulties on a numerical level. We obtain an acceptable solution in the case of a thin reaction zone, which occurs in $\eta = 1$ and has a thickness of 2ε. Thus, as an initial approximation in the vicinity of $\eta = 1$, we have $\eta^2 \dfrac{d^2\beta_T}{d\eta^2} \cong \dfrac{d^2\beta_T}{d\eta^2}$ and, if we multiply both sides of this equation by $(d\beta_T / d\eta)\, d\eta$:

$$\left[\frac{1}{2} \left(\frac{d\beta_T}{d\eta} \right)^2 \right]_{1-\varepsilon}^{1+\varepsilon} \cong -\Lambda \int_0^1 \omega(\beta_T)\, d\beta_T \qquad [3.23]$$

The constant Λ appears as an eigenvalue of the problem and gives us the flowrate in the permanent regime.

At the boundaries: in $\eta = 1 + \varepsilon : T = T_{ad},\ d\beta_T / d\eta = 0$ and in $\eta = 1 - \varepsilon$: $d\beta_T / d\eta = 1$. (When $\eta < 1 - \varepsilon$, we find ourselves dealing with a non-reactive zone with $d\beta_T / d\eta = const.$. The constant is equal to 1 since $\beta_T = 0$ in $\eta = 0$ and $\beta_T = 1$ in $\eta = 1$).

Thus, by inverting [3.23], we obtain

$$\Lambda = \left(2 \int_0^1 \omega(\beta_T)\, d\beta_T \right)^{-1} \qquad [3.24]$$

The integral is calculated by way of approximations. By taking, for $\omega(\beta_T)$, an expression of the form $(1-\beta_T)^{n-1} \beta_T^{\,p-1}$, we obtain the Rosen approximation:

$\int_0^1 \omega(\beta_T)\, d\beta_T = \Gamma(n)\Gamma(p) / \Gamma(n+p)$, valid for $T_a > 6\, T_{ad}$

Figure 3.5 gives the profiles of β and β_T for a laminar premixed flame. Given that we know Λ from equation [3.23], and given the first equation [3.22], we deduce the unitary mass flowrate: $\dot{m} = \left(\bar{\mathcal{M}}p/R\right)\sqrt{g\,BT^s/\Lambda}$ and the rate of combustion of the plane laminar premixed flame in relation to the fresh gases:

$$s_L^0 = \left(\bar{\mathcal{M}}p/\rho_f\, R\right)\sqrt{g\,BT^s/\Lambda} \qquad [3.25]$$

where ρ_f is the density of the fresh gases.

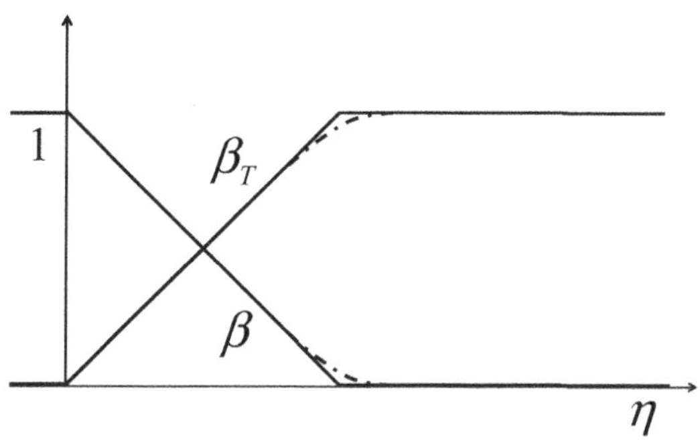

Figure 3.5. *Profiles of reduced temperature β_T and concentration β in a plane and laminar deflagration wave (premixed flame). Solid line: thin flame hypothesis. Dashed line: real-world flame*

3.1.3. *Other examples of a steady laminar premixed flame*

3.1.3.1. *Thin Bunsen burner flame*

The premixing occurs upstream from the burner outlet, which is usually a cylindrical tube. Figure 3.6 shows a cross-section of the thin flame front in a plane containing the axis of symmetry.

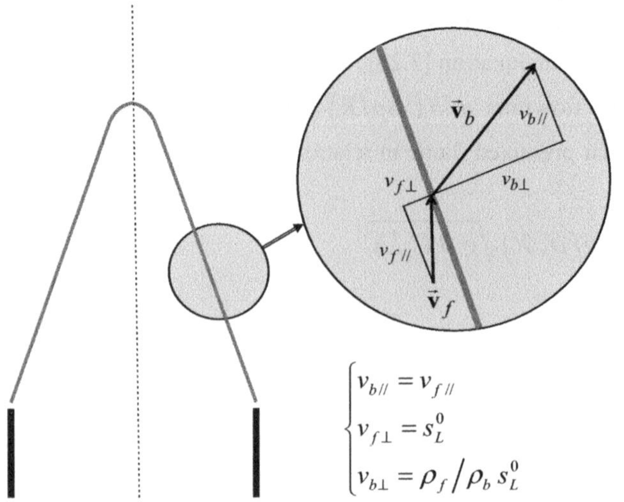

Figure 3.6. *Composition of the velocities on both sides of the oblique flame front produced by a Bunsen burner*

In reality, as was the case with the plane flame, the Bunsen burner flame has a certain thickness, as we were able to show in Figure 3.1.IV in [PRU 12], in the case of the methane/air mixture observed at low pressure.

In the rectilinear parts of the longitudinal section, relatively distant from the summit, the curvature of the flame is slight in comparison with its thickness and it behaves like a plane flame, i.e. like a plane discontinuity wave. The balance equations in Table 2.2 [PRU 12] are therefore valid. We can reasonably assume that in front the flame, the flow of fresh gases is steady and uniform with a velocity parallel to the axis of the burner and a density ρ_f. Behind it, the density of the burned gases ρ_b is necessarily less than ρ_f because of the heating resulting from combustion.

The conservation of mass is written as $[\rho \vec{\mathbf{v}}]_-^+ \cdot \vec{\mathbf{N}} = 0$, and results in the equality $\rho_f \, v_{fN} = \rho_b \, v_{bN}$, and earlier on we established $v_{fN} = s_L^0$, the velocity of the laminar plane flame, calculable on the basis of the thermodynamic and chemical kinetics data. The momentum equation is reduced to the conservation of the tangential velocity: $v_{fT} = v_{bT}$. With the interface notations used in Chapter 3 of [PRU 12], we would then have $\rho_f v_{f\perp} = \rho_b v_{b\perp}$, $v_{f\perp} = s_L^0$, $v_{f//} = v_{b//}$ (Figure 3.6). Further behind the flame, the trajectories are deviated by the flotation forces.

In addition, the situation is more complex at the summit of the flame, where the profile of the flame deviates from the cone (or dihedron in a 2D representation of the burner) and becomes rounded. Along the axis of symmetry, the normal velocity is v_f and is greater than the velocity of combustion $v_{fN} = s_L^0$ quoted above. The increase in combustion velocity from $s_L = s_L^0$ to $s_L = v_f$ is due to the curvature of the flame front, which stretches the flame as explained in [PRU 12] in section 3.2.4. Note that linear relations such as equation [3.10], averaging [3.11], are not valid with very great curvatures (see the works of Poinsot *et al.* [POI 92; POI 05] on this subject).

3.1.3.2. *Plane flame in front of a stagnation plane*

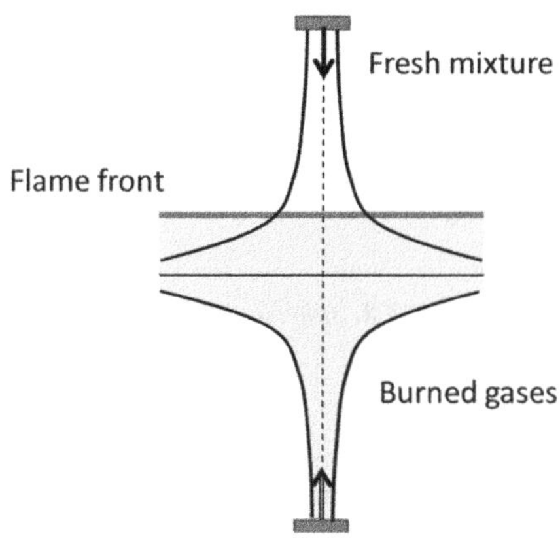

Figure 3.7. *Plane flame in front of a stagnation plane*

Consider two counter-current fluid flows, whose velocity fields are symmetrical in relation to the stagnation plane [POI 05]. These flows are steady and laminar. One of them comprises a dilute mixture of fresh gases in stoichiometric proportion. The other is formed of dilute burned gases stemming from the combustion of the former mixture re-injected. The energy involved in the combustion is therefore relatively low.

The flame front of the plane premixed flame is located above the stagnation plane. This time, we are dealing with a stretched flame and the speed of combustion s_L, which is different to s_L^0, is affected by the stretching rate of the flow.

3.1.3.3. *Spherical flame*

The fresh gases are initially distributed throughout the whole space, and are ignited at one point. We suppose that as the flame expands it retains spherical symmetry, with the burned gases remaining confined inside the sphere.

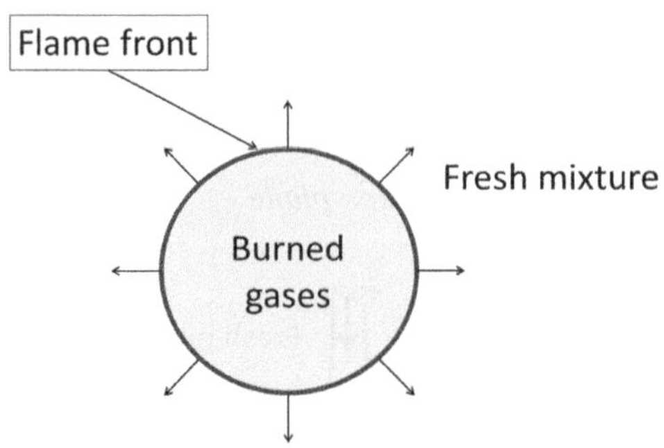

Figure 3.8. *Spherical flame*

3.1.3.4. *The G equation*

The G equation, introduced by Williams [WIL 85] applies to structures of thin flames propagating at a clearly-defined velocity of combustion.

In combustion, the state of the mixture depends on the velocity vector $\vec{\mathbf{v}}(\vec{\mathbf{x}},t)$, on the mass fractions Y_j of the species and on the temperature T, which are scalars. In a reactive medium, these scalars obey balance equations which include terms of chemical production and which are commonly called reactive scalars. If there is no chemical reaction, or if we consider the combinations of scalars which obey the equilibrium laws without chemical production such as $(\beta_i - \beta_j)$ or $(\beta_T - \beta_j)$ introduced in Chapter 2 of [PRU 12], we call these values non-reactive scalars or passive scalars.

Consider an isoscalar surface defined by an implicit equation (as given in section A1.3.11 in [PRU 12]): $F(\vec{\mathbf{x}},t)=G(\vec{\mathbf{x}},t)-G_0=0$. The mixture of fresh gases corresponds to $G(\vec{\mathbf{x}},t)<G_0$, and that of the burned gases to $G(\vec{\mathbf{x}},t)>G_0$. Thus, we can define the unitary normal oriented toward the burned gases as:

$$\vec{\mathbf{n}}=-\vec{\nabla}G/\left|\vec{\nabla}G\right| \quad [3.26]$$

If $\vec{\mathbf{v}}_f$ is the velocity of the fluid at the flame front and s_L the velocity of combustion in the normal direction, the velocity of the flame is then $\vec{\mathbf{w}}_f=\vec{\mathbf{v}}_f+s_L\vec{\mathbf{n}}$, and the value G satisfies the following balance equation, called the G equation:

$$\partial G/\partial t+\vec{\mathbf{v}}_f\cdot\vec{\nabla}G=s_L\left|\vec{\nabla}G\right| \quad [3.27]$$

3.2. Laminar non-premixed combustion

The most common example of a non-premixed flame is a *diffusion flame.* This name stems from the fact that it is diffusion which dictates the position of the flame. The reagents – the oxidizing and reducing agent – are first separated and then brought into contact, and combustion is generally initiated. The combustion products diffuse on both sides of the flame. For instance, if the chemical reaction is infinitely rapid, the position of the flame will be determined by writing that the concentration is stoichiometric. This is not a discontinuity wave as in the case of a premixed flame, whose flowrate was a value specific to the problem resulting from the coupling between the flow and the phenomena of diffusion and reaction.

An initial example of such a flame is the Burke–Schumann problem [BUR 28].

3.2.1. *Burke–Schumann problem*

Consider a diffusion flame between two coaxial flows of equal velocities – one of fuel and the other of oxidizer. The chemical reaction, supposed to be unique for the sake of simplicity, is of the form:

$$\nu'_F F+\nu'_O O\rightarrow \nu''_P P \quad [3.28]$$

We make the following hypotheses:

– validity of the Shvab-Zeldovich approximation presented in Chapter 2 of [PRU 12];

– at the exit from the coaxial tubes, the velocities of the fuel and oxidizer are equal and uniform. We can alter their proportion by modifying the diameters of the tubes and the concentrations;

– radial diffusion is negligible in comparison to axial diffusion: $\partial^2 Y_j / \partial r^2 << \partial^2 Y_j / \partial z^2$;

– mixing only occurs by diffusion and the radial velocity is zero: $v_r = 0$;

– the chemical reaction takes place when the proportion of the mixture is stoichiometric.

The equations for this problem are deduced from equation [2.205] of [PRU 12], where the definition of β_j is given by equation [2.204]. The reaction rate is eliminated by using the variable $\beta_F - \beta_O$. In cylindrical coordinates, the equation needing to be solved is

$$\frac{\partial \beta}{\partial z} - \frac{D}{r v_z} \frac{\partial}{\partial r}\left(r \frac{\partial \beta}{\partial r} \right) = 0 \qquad [3.29]$$

with the boundary conditions

$$\begin{cases} z = 0, r \in [0, r_F] \rightarrow \beta = -Y_F(r,0) / v'_F \mathcal{M}_F \\ z = 0, r \in [r_F, r_O] \rightarrow \beta = -Y_O(r,0) / v'_O \mathcal{M}_O \\ z > 0, r = 0, r_F \rightarrow \partial \beta / \partial r = 0 \end{cases}$$

Let us define the dimensionless coordinates: $\xi = r/r_0; \eta = Dz/r_0^{\,2} v_r$ and the reduced parameters: $C = r_F / r_0, \nu = \beta_{O0} / \beta_{F0}, \gamma = -\beta / \beta_{F0}$.

We shall use the variable separation method. The part of the solution which depends on ξ involves Bessel functions J_0 and J_1. We obtain (see [WIL 85; KUO 05] for details about the calculations):

$$\gamma = (1+\nu)C^2 - \nu + 2(1+\nu)C \sum_{n=1}^{\infty} \frac{1}{\phi_n} \frac{J_1(C\phi_n)}{\left[J_0(\phi_n)\right]^2} J_0(\phi_n \xi) e^{-\phi_n^{\,2}\eta} \qquad [3.30]$$

where the ϕ_n are the successive roots of the equation $J_1(\phi)=0$ (with the convention of order $\phi_n > \phi_{n-1}, \phi_0 = 0$).

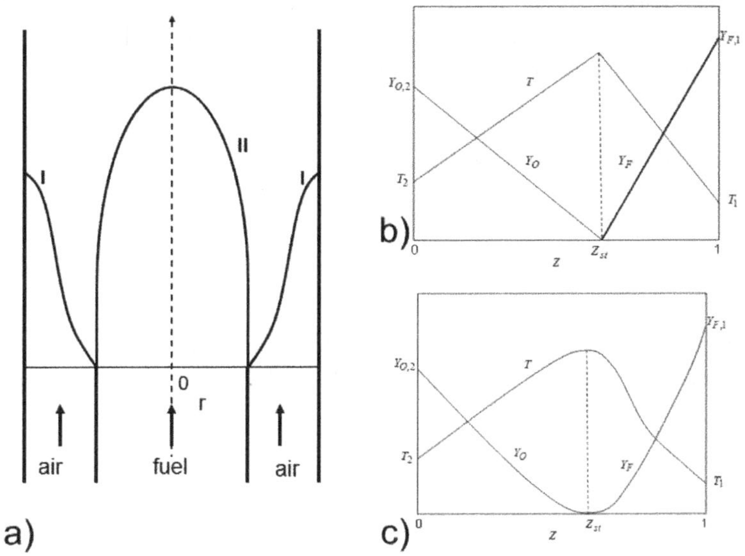

Figure 3.9. *a) Shapes of the diffusion flame for coaxial cylinders, I) under-ventilated, II) over-ventilated; b) Burke–Schumann-type Z solution; c) shape of a real Z solution*

We suppose that the reaction takes place at the surface of the flame (we also accept an infinitely rapid reaction giving a thin flame). The profile of the flame $\eta = f(\xi)$ can be deduced from the condition $\beta = 0$ and therefore $\gamma = 0$ in equation [3.30]. The height of the flame is then obtained by writing $\xi = 0$. If we overlook all the terms except $n = 1$, we obtain the following approximation for the dimensionless height $\eta(0) = f(0)$ of the under-ventilated flame:

$$\eta(0) = \frac{1}{\phi_1^2} \ln \left\{ \frac{\left[2(1+\nu)C\right] J_1(C\phi_1)}{\left[\nu - (1+\nu)\,C^2\right] \phi_1 J_0(\phi_1)} \right\} \qquad [3.31]$$

The height of the flames (obtained by limiting ourselves to $n = 1$) and the shapes of the flames[5] obtained by Burke and Schumann [BUR 28] (Figure 3.9a) are in close accordance with the experimental results.

Unlike what happens with premixed thin flames, the shape of the flame is found without using the energy balance equation. This equation should, however, be used if we wish to determine the field of temperature.

3.2.2. *Other examples of diffusion flames*

3.2.2.1. *The Z equation*

Let Z be the variable "mixture fraction".[6] It is a combination of the mass proportions of the fuel and oxidizer, and its stoichiometric value Z_{st} corresponds to the position of the flame, as in the Burke–Schumann problem discussed above. For the unique reaction [3.28], the variations in Y_F and Y_O due to the chemistry are $dY_F/\nu'_F\,\mathcal{M}_F = dY_O/\nu'_O\,\mathcal{M}_O$.

If we have two flows 1 and 2 with the mass flowrates $\dot{m}_1$ and $\dot{m}_2$, where the first flow contains the species F (mass fraction $Y_{F,1}$) and also a diluent and the combustion products, but not the species O, and the second flow contains the species O (mass fraction $Y_{O,2}$) and also a diluent and the combustion products, but not the species F, we have $Z = \dot{m}_1/(\dot{m}_1 + \dot{m}_2)$. If we define the stoichiometric ratio by $r_{st} = \nu'_O\,\mathcal{M}_O/\nu'_F\,\mathcal{M}_F$, we have

$$Z = \frac{r_{st}\,Y_F - Y_O + Y_{O,2}}{r_{st}\,Y_{F,1} + Y_{O,2}}$$

At the level of the flame, $r_{st}\,Y_F - Y_O = 0$ and $Z = \left(1 + r_{st}\,Y_{F,1}/Y_{O,2}\right)^{-1}$.

For a single binary diffusion coefficient, the mix fraction is a non-reactive scalar which obeys the conservative balance equation:

$$\rho\,\partial Z/\partial t + \rho\,\vec{v}\cdot\vec{\nabla}Z = \vec{\nabla}\cdot\left(\rho D\vec{\nabla}Z\right) \qquad [3.32]$$

---

5 The flame profiles obtained if we limit our reasoning to $n = 1$ are incorrect, and we need to further pursue the expansion in Bessel functions to obtain correct results (see section A7.1).

6 From the name Zeldovich.

which is called the "Z equation". Just like the G equation (equation [3.27]) with premixed combustion, this equation plays an important role in non-premixed turbulent combustion. Examples of a solution to a diffusion flame problem are shown in Figures 3.9b and 3.9c.

3.2.2.2. *Plane diffusion flame in the presence of a stagnation plane*

Figure 3.10 shows a counter-current diffusion flame with gaseous fuel and oxidizer. We suppose the diffusion coefficient to be constant. For dilute mixtures, such a flame is relatively simple to model because the density can be considered constant and the flame has no effect on the velocity field of the stretched flow under consideration [POI 05].

If y is the coordinate on the axis of flow (e.g. perpendicular to the flame), the components of the velocity of the fluid flow are $u = ax, v = -ay$ on the "flame side", which we assume *a priori* to be situated in the half-plane $y > 0$, as indicated in Figure 3.10.

Assuming Z to be independent of x and of z, the balance equation is written as $-ay\,\partial Z/\partial y = D\partial^2 Z/\partial y^2$. If we set $\eta = y\sqrt{a/D}$, we obtain the solution

$$Z = \frac{1}{2}\left[1 - \operatorname{erf}\left(\frac{\eta}{\sqrt{2}}\right)\right] \quad [3.33]$$

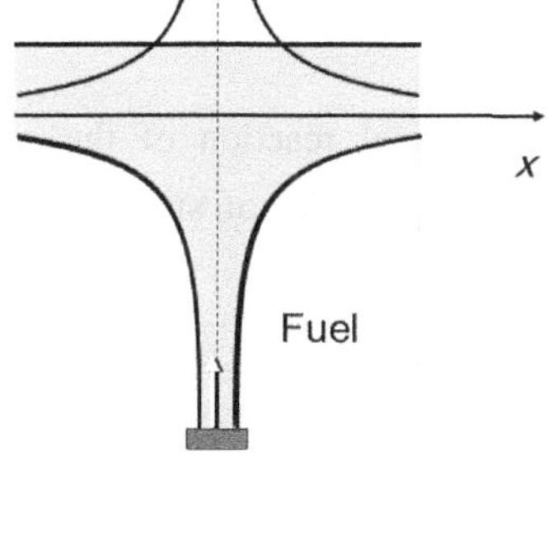

Figure 3.10. *An example of a stretched laminar diffusion flame*

The position of the flame is given by $Z_{st} = \frac{1}{2}\left[1 - \mathrm{erf}\left(\frac{\eta_f}{\sqrt{2}}\right)\right]$.

It is clear that for $Z_{st} < 1/2$, the flame is located on the oxidizer side of the stagnation plane (which is the case under investigation here), and conversely for $Z_{st} > 1/2$, the flame is on the fuel side of the stagnation plane.

When the changes in density occur, because of a release of heat, we also obtain a plane flame with a stretch rate imposed by the flow (but this is far more complicated to demonstrate, so we shall not attempt to here). The mix fraction assumes the same value as before (see equation [3.33]), but this time it is an approximate value [PET 00]:

$$Z = \frac{1}{2}\left(1 - \mathrm{erf}\left(\frac{\eta}{\sqrt{2}}\right)\right) \qquad [3.34]$$

and here, $\eta = \sqrt{a/D_\infty}\int_0^y \frac{\rho}{\rho_\infty} dy$, also assuming that $\rho^2 D = \rho_\infty{}^2 D_\infty$.

3.2.2.3. *Spherical diffusion flame around an evaporating drop*

The drop in question is spherical and comprises liquid fuel. It is initially surrounded by a gaseous oxidizer. We consider quasi-steady combustion of that drop, which we then suppose to be surrounded by a diffusion flame with spherical symmetry as well. By way of approximations, we wish to determine the regression law for the drop and calculate the position of the flame when it is thin.[7]

The main approximation used is the Shvab-Zeldovich approximation, presented in section 2.5.2 of [PRU 12] and which we have already used here. The theory dates from the 1950s and was simultaneously put forward by Godsave and by Spalding [GOD 53; SPA 53].

We assume a single chemical reaction of the same type as before, limiting ourselves to $F + O \rightarrow P$ for the sake of simplicity. The pressure is taken to be constant and uniform.

---

7 A similar problem is that of a drop of oxidizer (e.g. liquid oxygen) in a gaseous fuel (e.g. hydrogen).

The heat released by the flame causes the evaporation of the liquid fuel F in the drop, thereby releasing the fuel in gaseous form. This reacts with the gaseous oxidizer O in accordance with the reaction shown above.

The temperature of the drop is taken to be uniform and constant (this temperature is the saturating vapor temperature T_s). At infinity, the temperature is T_∞ and the mass fraction of the oxidizing agent is $Y_{O\infty}$ (Figure 3.11).

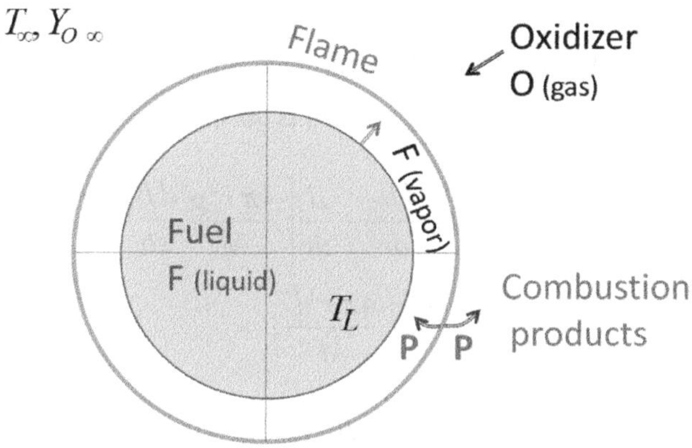

Figure 3.11. *Spherical diffusion flame*

If the density of the condensed phase is ρ_L and the radius of the drop is r_S, the mass flowrate of the gas through the sphere with radius r_S is given by $\dot{M} = -dM/dt = -4\pi r_s^2 \rho_L \, dr_s/dt$. Supposing this flowrate $\dot{M}$ to be linked to r_S, we can deduce the evolution of the radius of the droplet over time by simple integration. More specifically, the square of the diameter $d = 2r_S$ of the droplets is a linear function of time: $d^2 = d_0^2 - Kt$. This linear decrease of r^2 over time can also be observed in experiments.

Remember the following two equations obtained by the Shvab-Zeldovich approximation in section 2.5.2 of [PRU 12]:

$$\begin{cases} \vec{\nabla}\cdot\left(\rho\vec{\mathbf{v}}\beta_j - \rho D\vec{\nabla}\beta_j\right) = \dot{\zeta} \\ \vec{\nabla}\cdot\left(\rho\vec{\mathbf{v}}\beta_T - \lambda/c_{p,f}\,\vec{\nabla}\beta_T\right) = \dot{\zeta} \end{cases} \qquad [3.35]$$

$\beta_j = Y_j/\nu_j\mathcal{M}_j$ is a concentration variable and

$$\beta_T = \sum_j Y_j \int_{T^0}^{T} c_{p,j}\, dT \Big/ \Delta H$$

is a temperature variable (equations [2.204] and [2.212]). We suppose a Lewis number $Le = \lambda/\rho D c_{pf}$ to be equal to 1. We can therefore set $g = \lambda/c_{pf} = \rho D$ and take that g to be constant. The following system is obtained:

$$\begin{cases} 4\pi r^2\,\rho v = \dot{M} \cong const., \dfrac{d\beta_j}{dr} - \dfrac{d}{dr}\left[\dfrac{4\pi r^2 g}{\dot{M}}\dfrac{d\beta_j}{dr}\right] = \dfrac{4\pi r^2\dot{\zeta}}{\dot{M}}, \\ \dfrac{d\beta_T}{dr} - \dfrac{d}{dr}\left[\dfrac{4\pi r^2 g}{\dot{M}}\dfrac{d\beta_T}{dr}\right] = \dfrac{4\pi r^2\dot{\zeta}}{\dot{M}}, \\ -g\left.\dfrac{d\beta_j}{dr}\right|_{r=r_s} = -(\rho v)_L\left(\beta_{js} - \beta_{jL}\right), -g\left.\dfrac{d\beta_T}{dr}\right|_{r=r_s} = -\dfrac{(\rho v)_L\, l}{\Delta H} \end{cases} \qquad [3.36]$$

By setting $\xi = \dot{M}/4\pi g r$, we find $d^2\left(\beta_T - \beta_O\right)/d\xi^2 + d\left(\beta_T - \beta_O\right)/d\xi = 0$, an equation whose integration gives us the formula $\beta_T - \beta_O = A' + B'e^{-\xi}$, where the constants A' and B' can be determined by looking at the boundary conditions.

We obtain $A' = \beta_{Ts} + l/\Delta H$, $B' = -\left(\beta_{O\infty} + \beta_{Ts} + l/\Delta H\right)$ and

$\left(l/\Delta H\right)e^{\xi_s} = \beta_{O\infty} + \beta_{Ts} + l/\Delta H$, which gives us the result

$$\dot{M} = 4\pi g r_s \ln\left[\left(\beta_{O\infty} + \beta_{Ts}\right)\Delta H/l + 1\right] \qquad [3.37]$$

As indicated above, as the mass flowrate is proportional to the radius of the droplets, this gives us the law in $d^2 = d_0^{\,2} - Kt$, and here

$$K = \left(8g/\rho_L\right)\ln\left[\left(\beta_{O\infty} + \beta_{Ts}\right)\Delta H/l + 1\right] \qquad [3.38]$$

The quantity $(\beta_T - \beta_O)$ evolves in accordance with the rule

$$\beta_T - \beta_O = \beta_{Ts} + l/\Delta H - (\beta_{O\infty} + \beta_{Ts} + l/\Delta H) e^{-\xi} \qquad [3.39]$$

Let us now study $(\beta_F - \beta_O)$. In particular, this will give us insight into the concentrations and if the flame is thin and the combustion is stoichiometric, $\beta_F = \beta_O$ will give us its position. Of course, we find that $\beta_T - \beta_O = A + B e^{-\xi}$, the constants A and B being deduced from the boundary conditions at infinity and at the surface of the drop: $A = -1/\mathcal{M}_F$, $B = 1/\mathcal{M}_F - \beta_{O\infty}$ and the mass fraction Y_{Fs} at the surface of the drop is deduced from $\beta_{F\,s} = 1/\mathcal{M}_F + (1/\mathcal{M}_F - \beta_{O\infty}) e^{-\xi_s}$.

The position of the *spherical thin flame* is deduced from these results. We obtain

$$\frac{r_f}{r_s} = \frac{\ln\left[1 + (\beta_{O\infty} + \beta_{Ts}) \Delta H / l\right]}{\ln(1 - \mathcal{M}_F\, \beta_{O\infty})} \qquad [3.40]$$

The concentration and temperature profile are obtained as follows in the case of a thin flame.

For $r_s < r < r_f$, $\beta_O = 0$ and $\dot{\zeta} = 0$, then $\beta_F = A + B e^{-\xi}$, $\beta_T = A' + B' e^{-\xi}$, the four coefficients having been determined previously.

When $r_f < r$, we always have $\dot{\zeta} = 0$ but this time $\beta_F = 0$. Thus, $\beta_O = -A - B e^{-\xi}$, $\beta_T = A' - A + (B' - B) e^{-\xi}$.

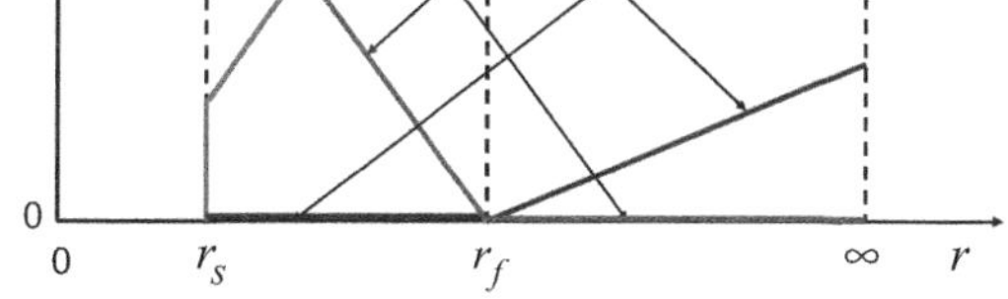

Figure 3.12. *Profiles of temperature and mass fractions of a spherical droplet in combustion on the scale of* $e^{-\xi}$ *with* $\xi = \dot{M}/4\pi g r$, $g = \lambda/c_{pf} = \rho D$

We can see that, with the hypothesis of a thin flame and with an infinitely rapid chemical reaction, the evolutions are governed by the processes of convection and diffusion. The chemical reaction only plays a role because of the energy that it provides.

The same is not true with a premixed flame – surrounding a droplet of mixture, for instance – because the rate of combustion will also depend on the coupling with the chemical kinetics, as we saw in our study of deflagration waves.

The theoretical profiles can be represented as a function of r on the scale $e^{-\xi}$ with $\xi = \dot{M}/4\pi g r$ (Figure 3.12).

The results of this theory were compared with those of the experiments done in microgravity by Kumagai [KUM 70] (also see [PRU 88]): there may well be relatively close correspondence with the time-dependent linear regression law found theoretically and governing the square of the drop's radius. However, the same is not true with regard to the ratio r_f/r_s, the radius of the flame to that of the droplet. Equation [3.40] indicates that the ratio r_f/r_s is constant, but the experimental results show otherwise.

More complex calculations which take account of the unsteady characteristics of the process and of the variations in the transfer coefficients with changing temperature [NAJ 85; SCH 85] yield results which correspond more closely to the experimental results. Note that the values of r_s and r_f/r_s have a significant influence on the performances of a rocket engine, in which the flow is made up of condensed burning drops. The evolution of the radius of the droplet over time gives the combustion time and the length of the combustion chamber. The ratio r_f/r_s indicates the possible interactions between the drops during the course of combustion in the flow. The previous theory applies only in the absence of interactions. The effects of coupling between the flow and the individual process of combustion that each drop causes are much more difficult to evaluate [BOR 95; DIE 99].

3.3. Turbulent combustion

Turbulence has a significant impact on the combustion process.[8] Of course, there is a distinction to be drawn between turbulent premixed flames and turbulent diffusion

8 The author wishes to thank Nicolas Bertier for his valuable feedback, which helped improve and enrich section 3.3 and Appendix 3.

flames. However, another distinction can also be made, in relation to the rapidity of the chemical processes, between infinitely rapid chemistry and chemistry with a finite reaction rate [DUP 86].

Now having presented the general form of the averaged balance equation, we shall give the Peters diagrams defining the different combustion regimes. The averaged equations bring additional terms into play in relation to the case of a non-reactive simple fluid[9] as presented in Appendix 3; the closure of the species- and energy-balance equations is a key problem in terms of turbulent combustion, and we shall give a review of some models of combustion that can be used to solve these equations. Finally, we shall touch on the Large Eddy Simulation (LES) in combustion.

3.3.1. *Averaged balance equation for turbulent combustion*

One opportunity for studying turbulent combustion is being able to separate the equations relating to the velocity field from those relating to the scalar quantities of temperature and mass fractions.

Let us apply the general balance equation [A3.32] $\frac{\partial(\rho f)}{\partial t}+\vec{\nabla}\cdot\left(\rho f\,\vec{\mathbf{v}}+\vec{\jmath}_F\right)=\dot{W}_F$ to a scalar property such as the mass fraction Y_j or the temperature T, and suppose that the flux $\vec{\jmath}_F$ involves only one diffusion coefficient D.

The balance equation becomes $\rho\frac{\partial f}{\partial t}+\rho\,\vec{\mathbf{v}}\cdot\vec{\nabla}f=\vec{\nabla}\cdot\left(\rho D\vec{\nabla}f\right)+\dot{W}_F$. As we do not know the density *a priori*, we shall use the Favre average $\tilde{f}$ such that $f=\tilde{f}+f''$. Thus, by setting $\dot{W}_F=\rho S_F$, we obtain

$$\bar{\rho}\frac{\partial\tilde{f}}{\partial t}+\bar{\rho}\,\tilde{\mathbf{v}}\cdot\vec{\nabla}\tilde{f}=\vec{\nabla}\cdot\overline{\left(\rho D\nabla f\right)}-\nabla\cdot\left(\bar{\rho}\,\widetilde{\mathbf{v}''f''}\right)+\bar{\rho}\,\widetilde{S_F}\qquad[3.41]$$

9 In the sense of a "fluid with a single component" as described in Chapter 1 of [PRU 12].

With a large Reynolds number, as is the case with a turbulent flow, the term of molecular diffusion $\vec{\nabla}\cdot\overline{(\rho D\nabla f)}$ is negligible. In addition, the source term, which is proportional to $S_T(T)=B(T_b-T)\exp(-E_a/RT)=S_T(\tilde{T})$ $\left[1-T''/(T_b-T)\right]\exp(E_a)T''/R\tilde{T}^2$, gives an increase in the fluctuations of chemical production around its average value.

For non-reactive scalars, we can write $\widetilde{\mathbf{v}''f''}=-D_t\nabla\tilde{f}$ [PET 00].

The equation governing the squares of the fluctuations $\widetilde{f''^2}$, for its part, is written as

$$\bar{\rho}\frac{\partial\widetilde{f''^2}}{\partial t}+\bar{\rho}\tilde{\mathbf{v}}\cdot\vec{\nabla}\widetilde{f''^2}=-\nabla\cdot\left(\bar{\rho}\widetilde{\mathbf{v}''f''^2}\right)-2\bar{\rho}\widetilde{\mathbf{v}''f''}\cdot\nabla\tilde{f}-\bar{\rho}\tilde{\chi}_F+2\bar{\rho}\widetilde{f''S''}_F$$

[3.42]

where the average dissipation rate associated with the scalar F is $\tilde{\chi}_F=2D\left(\widetilde{\nabla f''}\right)^2$.

We define a scalar integral time-scale: $\tau_F=\widetilde{f''^2}/\chi_F$. In the *non-reactive case*, τ_F is proportional to time $\tau=\tilde{k}/\tilde{\varepsilon}\Rightarrow\tau=c_\chi\tau_F$, where $c_\chi\approx 2\Rightarrow\chi_F=c_\chi\tilde{\varepsilon}/\tilde{k}\,\widetilde{f''^2}$. As happens with a steady inertial cascade regime, there is equality between production and dissipation, so

$$-2\bar{\rho}\widetilde{\mathbf{v}''f''}\cdot\nabla\tilde{f}=\bar{\rho}\tilde{\chi}_F\Rightarrow-2\,\widetilde{\mathbf{v}''f''}\cdot\nabla\tilde{f}=c_\chi\left(\tilde{\varepsilon}/\tilde{k}\right)\widetilde{f''^2}.$$

By multiplying both sides of this equation by $D_t\propto\tilde{k}^2/\tilde{\varepsilon}$, we obtain

$$-D_t\widetilde{\mathbf{v}''f''}\cdot\nabla\tilde{f}\approx c_\chi\tilde{k}\,\widetilde{f''^2}$$

Assuming there is isotropy, and therefore a relation of proportionality between the two quantities, we deduce that the hypothesis of the turbulent in gradient law is valid:

$$\widetilde{\mathbf{v}''f''} \approx -c_\chi^{-1} D_t \nabla \tilde{f} \qquad [3.43]$$

We shall accept that this linear law remains true in the case of reactive scalars [PET 00].

3.3.2. *Premixed turbulent combustion regimes*

The classification in Figure 3.13, drawn from [PET 00], relates to premixed flames. This classification assumes that the turbulence is in the inertial zone of the turbulent energy spectrum described in Appendix 3 and the flames are thin, i.e. that they have the structures of a deflagration wave with a high activation energy.

In Figure 3.13, the separation lines between the domains are defined on the basis of the characteristic values and dimensionless numbers. Some of these values are related to conventional turbulence, such as:

– the integral scale ℓ defined in Appendix 3;

– the intensity of turbulence $v' = \sqrt{\overline{u'^2}}$;

– the Kolmogorov length scale ℓ_K .

Other values relate to combustion:

– the thickness of preheating of the flame $\ell_f = D/s_L$; for reasons of scale, it is supposed that there is a single diffusion coefficient for all the species and that the Schmidt number Sc = ν / D is equal to 1;

– the thickness of reaction of the flame $\ell_\delta = \delta \ell_f$; for a chemical reaction with high activation energy, we have $\delta << 1$;

– the normal speed of combustion s_L, also deduced from laminar deflagration wave theory;

– the Gibson scale $\ell_G = s_L^3/\varepsilon$ is a scale at which the turbulence can affect the flame front.

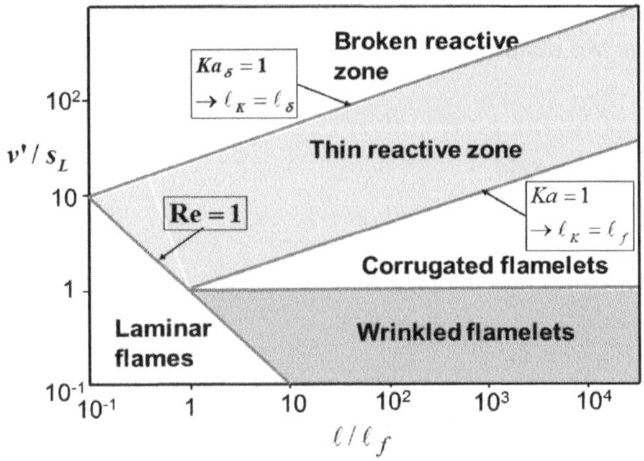

Figure 3.13. *Peters diagram [PET 00] for premixed combustion*

The dimensionless numbers are:

– the turbulent Reynolds number $\text{Re} = v'\ell/s_L\ell_f$, relating to the thickness of the flame;

– the turbulent Damköhler number $Da = s_L\ell/v'\ell_f$;

– the first Karlovitz number $Ka = \ell_f^{\,2}/\ell_K^{\,2}$; we have $\text{Re} = Da^2Ka^2$;

– the second Karlovitz number $Ka_\delta = \ell_\delta^{\,2}/\ell_K^{\,2}$.

We shall now describe the different regimes in Figure 3.13. Just before we do so, it is important to explain the concept of flamelets, introduced by Williams [WIL 75], which is now used in all types of turbulent combustion (we should also mention Peters, Kuznetsov and Bray for their extensions of this concept). Flamelets are thin reactive-diffusive layers integrated within a field of non-reactive turbulent flow. Poinsot and Veynante [POI 05] give a less restrictive definition of the flamelets regime which corresponds to a continuous flame front without extinction[10], and where no reference is made to thin flames.

10 Extinguishing of the flame occurs when a flame front is subjected to external disturbances such as heat losses or aerodynamic stretching of sufficient magnitude to reduce the reaction rate to a negligible value and, in some cases, completely eliminate the combustion process. In certain cases, the accumulation of combustion products in a confined space may be a cause of extinguishing (Appendix 6).

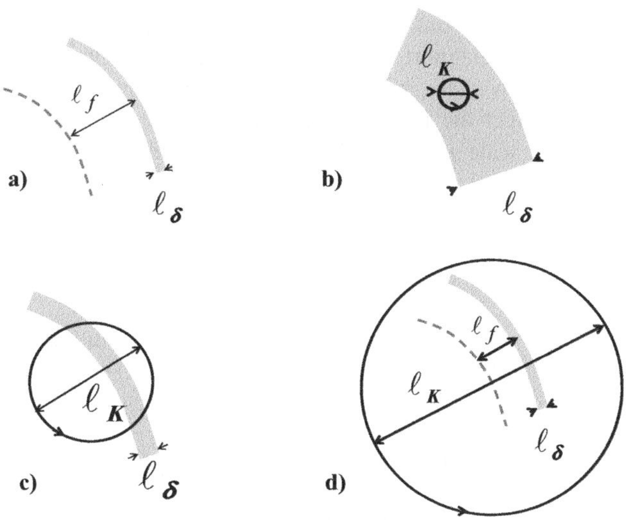

Figure 3.14. *Premixed turbulent combustion regimes. a) Flame structure; b) small scales of turbulence ℓ_K less than the thickness of the reaction zone; c) $\ell_\delta < \ell_K < \ell_f$; d) $\ell_\delta < \ell_f < \ell_K$*

The laminar domain is limited to the region where Re << 1. The other regions of the diagram can be explained by comparing the scales defined above – particularly the Kolmogorov length scale ℓ_K with the characteristic thicknesses of the premixed flamelet, as can be seen in Figure 3.14. In cases (b) and (c), the reaction zone can be considered to be a perfectly mixed homogeneous reactor (see Figure 2.8 in Chapter 2): combustion is initiated when the upper branch is reached by the effect of the heat; extinction can occur if the heat injection is too small.

In case (d), the zone of corrugated flamelets is distinguished from the zone of wrinkled flamelets using the Gibson scale, where the speed of the flame s_L^0 is equal to the turbulent velocity v' (see [PET 00]). For $v' < s_L$, the form of the flamelets is relatively undisturbed: the velocity of the turbulent motion is too slow to sufficiently wrinkle the flame front and thereby facilitate interactions of the flames. For $v' > s_L$, the turbulent motions are able to rumple this flame front until interactions occur between the flames, leading to the formation of pockets of fresh and burned gases; this is the "thin flame with pockets regime" or "corrugated flamelets regime".

Certain lacunas in the above classification of regimes of premixed flames were pointed out by Poinsot and Veynante, who used the results of direct numerical simulations (DNS) and experimental results to justify the modification of the combustion diagrams. They put forward numerous arguments in favor of their improvements to Peters' diagram (see [POI 05; PRU 10]). One of the arguments related to the Kolmogorov scale, which was deemed to be too small with too low a velocity to properly affect the flame front; it would, they said, be advantageous to replace the Kolmogorov scale with the Gibson scale defined above.

The modified combustion diagrams take account of these considerations and of heat loss due to radiation, the significant stretching of the flame under the influence of pairs of eddies, etc. They can be consulted in [POI 05].

An example of a wrinkled flame is given in [PRU 10] with a premixed Bunsen burner flame subjected to the influence of gravity.

3.3.3. *Non-premixed turbulent combustion regimes*

It is more complicated to determine the scales in the case of non-premixed combustion than in the case of premixed combustion. Indeed, there is no simple definition for the thickness of the flame such as given in section 3.3.2, and the turbulent scales are evaluated partially in the space of the mixture fraction Z introduced in section 3.2.2.1.

We have:

– a flame thickness $\ell_D = \sqrt{D_{st}/a}$ defined on the basis of the square root of the ratio of the diffusion coefficient, taken at the position of the flame, when $Z = Z_{st}$, to the stretch rate a;

– the corresponding thickness of diffusion in the mixture fraction space: $\Delta Z_f = \ell_D \left|\nabla Z\right|_{st}$. This thickness of diffusion can be rewritten as a function of the stretch rate a and an instantaneous value of the local scalar dissipation rate in equation [3.42] for a stoichiometric mixture $\chi_{st} = 2D_{st}\left|\nabla Z\right|_{st}^{2}$. Thus, we have $\Delta Z_f = \sqrt{\chi_{st}/2a}$;

– a thickness of reaction $\Delta Z_R = \varepsilon \Delta Z_f$ equal to the thickness of the oxidation layer, which is proportional to $\chi_{st}^{1/4}$ for a methane–air diffusion flame in four stages [PET 00]:

$$
\begin{array}{ll}
I & CH_4 + 2H + H_2O = CO + 4H_2 \\
II & CO + H_2O = CO_2 + H_2 \\
III & H + H + M = H_2 + M \\
IV & O_2 + 3H_2 = 2H + 2H_2O
\end{array}
$$

Finally, we obtain $\Delta Z_R = \varepsilon_q \left(\chi_{st}/\chi_q\right)^{1/4} \Delta Z_f$, where ε_q and χ_q are the values which correspond to the extinction boundary.

For turbulent diffusion flames, ΔZ_f and ΔZ_R need to be compared to the fluctuation of intensity of the mixture fraction $Z' = \left(\widetilde{Z''^2}\right)^{1/2}$, and more specifically to $Z'_{st} = \left(\widetilde{Z''^2_{st}}\right)^{1/2}$ at the position of the flame where $\tilde{Z}(\mathbf{x},t) = Z_{st}$.

The coordinates chosen for the diagram in Figure 3.15 [PET 00] are therefore $\left(\chi_{st}/\chi_q\right)$ and $\left(Z'_{st}/\Delta Z_f\right)$.

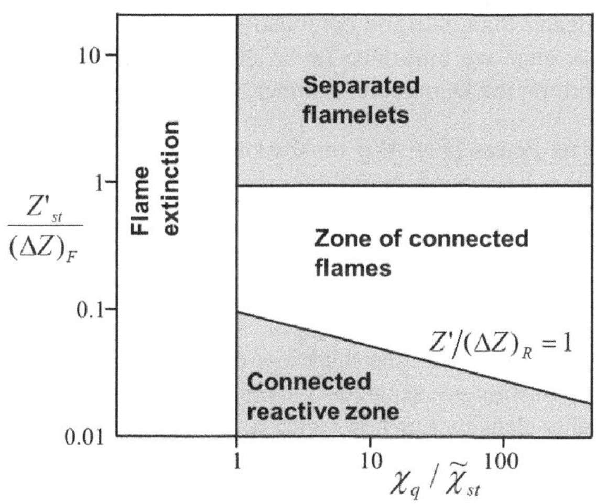

Figure 3.15. *Simplified representation of the Peters diagram [PET 00] for non-premixed turbulent combustion regimes*

We observe four regions in this diagram:

– a zone of extinction of the flame for $\chi_q / \tilde{\chi}_{st} < 1$;

– a zone of separate flamelets for $Z'_{st} / \Delta Z_f > 1$;

– two connected zones of combustion, separated by a line with gradient (-1/4) in logarithmic coordinates, which corresponds to $Z'_{st} = \Delta Z_f$. For $Z'_{st} > \Delta Z_f$, the fluctuations of the mixture fraction are sufficient to separate the diffusion zones surrounding the reaction zone; for $Z'_{st} < \Delta Z_f$, the fluctuations of the mixture fraction are slight and mixing is intense.

3.3.4. *Models of turbulent combustion*

Numerous attempts have been made to obtain closure relations for the turbulent combustion equations. This is due in particular to the large number of cases which arise, and also to the nonlinearity of the source terms.

For "infinitely rapid" chemistry, the Damköhler parameter $Da = \tau_t / \tau_{\text{chem}}$ (defined on the basis of a characteristic time of turbulence τ_t and the chemical time τ_{chem}), is much greater than one, and combustion is controlled only by the turbulent mixture. However, once we introduce finite chemical kinetics, the composition of the mixture depends on the Damköhler number.

Authors such as Peters [PET 00] on the one hand and Poinsot and Veynante [POI 05] on the other have put forward the main models for combustion. Below, we shall examine a few of these.

3.3.4.1. *Bray-Moss-Libby model*

The Bray-Moss-Libby (BML) model applies to a premixed flame with very high reaction activation energy. Thus, the thickness of the flame is infinitely slight, and turbulence and combustion are separate. This model combines a statistical approach, using the probability density function, with a physical analysis of the flamelets' regime.

Consider a progress variable, which might be $c = (T - T_f)/(T_b - T_f)$ or $c = Y_P / Y_{Pb}$; we accept that the normalized probability density function of c is defined by two Dirac distributions:

$$P(c;\mathbf{x},t) = \alpha(\mathbf{x},t)\delta(c) + \beta(\mathbf{x},t)\delta(1-c), \text{ where } \int_0^1 P(c;\mathbf{x},t)\,dc = 1 \qquad [3.44]$$

and we have $\alpha(\mathbf{x},t)+\beta(\mathbf{x},t)=1$. As the pressure is constant and uniform as a first approximation, and the average molar mass is the same for the fresh and burned gases, we can write: $\dfrac{\rho}{\rho_f}=\dfrac{T_f}{T}=\dfrac{1-\gamma}{1-\gamma(1-c)}$, where $\gamma=1-\dfrac{\rho_b}{\rho_f}$.

The following results enable us to make deductions about the average values of ρ and c:

$$\begin{cases}\bar{\rho}/\rho_f=\int_0^1(\rho/\rho_f)\,P(c;\mathbf{x},t)\,dc=1-\gamma\beta(\mathbf{x},t)\\ \tilde{c}=\overline{\rho c}/\bar{\rho}=(\rho_f/\bar{\rho})\int_0^1\rho c\,P(c;\mathbf{x},t)\,dc=\beta(1-\gamma)/(1-\gamma\beta(\mathbf{x},t))\\ \bar{c}=\int_0^1 c\,P(c;\mathbf{x},t)\,dc=\tilde{c}/(1-\gamma)(1-\tilde{c}),\quad \bar{\rho}/\rho_f=(1-\gamma)/[1-\gamma(1-\tilde{c})]\end{cases}\qquad [3.45]$$

We can also envisage a velocity component u in addition to c. We now have

$$\begin{cases}P(u,c;\mathbf{x},t)=\alpha(\mathbf{x},t)\delta(c)P(u_f;\mathbf{x},t)+\beta(\mathbf{x},t)\delta(1-c)P(u_b;\mathbf{x},t),\\ \text{where}\int_0^1 P(u,c;\mathbf{x},t)\,du\,dc=1,\ \tilde{u}(\mathbf{x},t)=(1-\tilde{c})\bar{u}_f(\mathbf{x},t)+\tilde{c}\bar{u}_b(\mathbf{x},t)\end{cases}\qquad [3.46]$$

For the Favre correlation, $\widetilde{u''c''}=\overline{\rho(u-\tilde{u})(c-\tilde{c})}/\bar{\rho}$, we obtain the relation

$$\widetilde{u''c''}=\tilde{c}(1-\tilde{c})(u_b-u_f)\qquad [3.47]$$

This relation can be compared to equation [A3.37], obtained by the classic application of the gradient theory to the Favre correlation (section A3.3.1) and to equation [3.43], which can be written as

$$\widetilde{u''c''}=-D_t\,\partial\tilde{c}/\partial x\qquad [3.48]$$

Note that, to obtain a steady plane turbulent flame, there is a contradiction between [3.47], which is resolved experimentally and [3.48], where the gradient $\partial c/\partial x$ takes the place of the jump $(1-\tilde{c})$. Indeed, as $u_b > u_f$, equation [3.47] tells us that $\widetilde{u''c''}>0$, which gives a negative concentration gradient in equation [3.48]. This result is referred to as *counter-gradient diffusion*. In fact, it is a combination of two phenomena: the turbulent mixing and the expansion of the gas. Counter-gradient diffusion is due to the expansion of the gas at the flame front (see [VEY 97] for a more detailed analysis of this problem). We now need to calculate the turbulent source term resulting from the chemical reaction, which appears in the averaged balance equation of c:

$$\bar{\rho}\left(\partial\tilde{c}/\partial t+\mathbf{v}\cdot\nabla\tilde{c}\right)+\nabla\cdot\left(\bar{\rho}\widetilde{\mathbf{v}''c''}\right)=\tilde{\dot{W}}_c \qquad [3.49]$$

The rate of scalar dissipation $\tilde{\dot{W}}_c$ is determined using the notion of frequency of flame crossing, or using flame surface density models [PET 00; POI 05].

3.3.4.2. *Averaged G equation*

We introduced the G equation for laminar premixed flames in section 3.1.3.4.

It is practical to use this equation, for instance, when we are studying the regime of corrugated flamelets [PET 00].[11]

We shall now define the probability density function $P(G;\mathbf{x},t)$ and also the Favre average of G and its variance:

$$\bar{\rho}\tilde{G}=\int_{-\infty}^{+\infty}\rho G P(G;\mathbf{x},t)dG,\ \bar{\rho}\widetilde{G''^2}=\int_{-\infty}^{+\infty}\rho G''^2 P(G;\mathbf{x},t)dG \qquad [3.50]$$

After analyzing the different terms in these equations and making approximations in order to establish closure relations, Peters obtains the following system:

$$\begin{cases}\bar{\rho}\left(\partial\tilde{G}/\partial t+\mathbf{v}\cdot\nabla\tilde{G}\right)=\bar{\rho}s_T^0\left|\nabla\tilde{G}\right|-\bar{\rho}D_t\tilde{\kappa}\left|\nabla\tilde{G}\right| \\ \bar{\rho}\left(\partial\widetilde{G''^2}/\partial t+\mathbf{v}\cdot\nabla\widetilde{G''^2}\right)=\nabla_{//}\cdot\left(\bar{\rho}D_t\nabla\widetilde{G''^2}\right)+2\bar{\rho}D_t\left(\nabla\tilde{G}\right)^2 \\ -c_s\bar{\rho}\left(\tilde{\varepsilon}/\tilde{k}\right)\widetilde{G''^2}\end{cases} \qquad [3.51]$$

where s_T^0 is the speed of turbulent combustion, $\tilde{\kappa}$ the average curvature of the flame front, D_t the turbulent diffusivity and c_s is a modeling constant which can be taken as being equal to 2. The rate of turbulent combustion is not a clearly defined

11 Peters calls this the "level set approach".

quantity[12], and depends particularly on the combustion regime (for further information see [PET 00]). Additional developments have been made to the approach to deal, in particular, with problems of spark ignition [LIA 06].

3.3.4.3. *Eddy break-up models*

In an eddy break-up (EBU) model, the chemistry plays no explicit role; the turbulent motions control the reaction rate. The reaction zone is considered to be a collection of pockets of fresh and burned gases, transported by turbulent eddies.

This model was put forward by Spalding [SPA 73] in association with the model $k-\varepsilon$ used to describe turbulence. The average rate of fuel consumption F in the turbulent premixed flame is expressed as follows:

$$\tilde{\dot{W}}_F = -C_{EBU}\bar{\rho}\frac{\varepsilon}{k}\sqrt{\widetilde{Y''^2_F}} \qquad [3.52]$$

where C_{EBU} is a constant which is adjusted as a function of the mixture in question.

For diffusion flames, we model:

$$\sqrt{\widetilde{Y''^2_F}} = \min\left(\tilde{Y}_F, \tilde{Y}_O / r_{st}\right)$$

where the stoichiometric ratio is defined by $r_{st} = \nu_O \mathcal{M}_O / \nu_F \mathcal{M}_F$.

For premixed flames, we can use any of the reduced variables defined in section 3.3.4.1 such that $c = (T - T_f) / (T_b - T_f)$, instead of Y_F. The corresponding average production rate is then $\tilde{\dot{W}}_F = -C_{EBU}\bar{\rho}\frac{\varepsilon}{k}\sqrt{\widetilde{c''^2}}$.

The estimation of the fluctuation $\widetilde{c''^2}$ for an infinitely thin flame yields

12 Damköhler was the first to present the theoretical expressions for the rate of turbulent combustion [DAM 40]. Beginning with the equivalence between the mass flux of a laminar flow across the flame surface A_T and the mass flux of the turbulent flow through the normal cross-section A, he deduced the following relation between the rates of laminar and turbulent combustion of a steady premixed flame in a tube: $\dot{m} = \rho_f s_L A_T = \bar{\rho}_f s_T A$.

$$\widetilde{c''^2} = \overline{\rho\left(c-\tilde{c}\right)^2} = \overline{\rho}\left(\widetilde{c^2} - \tilde{c}^2\right) = \overline{\rho}\,\tilde{c}\left(1-\tilde{c}\right),$$

because the temperature can only assume one of the two values $c = c_f = 0$ or $c = c_b = 1$, and the average reaction rate is therefore $\tilde{\dot{W}}_F = -C_{EBU}\overline{\rho}\frac{\varepsilon}{k}\tilde{c}\left(1-\tilde{c}\right)$ [POI 05].

This model, which has been compared to the Arrhenius model, is sometimes used in conjunction with it in industrial codes. Certain adjustments to the constant C_{EBU} have been suggested, with a view to integrating certain functionalities [SAI 88].

3.3.4.4. *Passive scalar models*

By combining the balance equations to eliminate the terms of chemical production (see section 2.5.2 of [PRU 12]), we introduce a parameter $\phi = r_{st}Y_F - Y_O$ (called the Shvab-Zeldovich variable), or the corresponding reduced variable $Z = \dfrac{\phi - \phi_{\min}}{\phi_{\max} - \phi_{\min}}$. We can work back to the concentrations by using the notion of the probability density function:

$$\tilde{Y}_F = \int_0^1 Y_F(\phi)P(\phi)d\phi \qquad [3.53]$$

Models derived from this one include the CRAMER model [DUP 86] and an extension to it, the CLE model [RAV 97].

3.3.4.5. *Coherent flame model*

The coherent flame model uses the concept of flame stretching (see section 3.2.4 of [PRU 12]) such as it is applied to the density of area of the flame [LIB 80]. Marble and Broadwell are responsible for this model, which has been adapted to diffusion flames [CAN 90]. It is called the "flame surface density model" in [POI 05]. Let us write the production term from equation [3.49], which is valid for a thin premixed flame, as in [PET 00]: $\tilde{\dot{W}}_c = \rho_f s_L^0 I_0 \Sigma$, where ρ_f is the density of the fresh gases, s_L^0 is the speed of laminar combustion, Σ is the flame surface density (the surface area of the flame per unit volume), and I_0 is a stretch factor which can be deduced from the formula $I_0 = \left(1/s_L^0\right)\int_0^\infty s_c(\kappa)p(\kappa)d\kappa$ [POI 05].

The stretch factor is of order 1, s_c is the speed of consumption, κ is the stretch rate, and the probability $p(\kappa)$ is supposed (in most practical cases) to be a Dirac function $p(\kappa)=\delta(\kappa-\bar{\kappa})$.

An acceptable formulation (which yields satisfactory results in relation to the DNS results) for the surface density balance equation is [TRO 94] equation [3.12] in [PRU 12].

3.3.4.6. *Arrhenius model*

The simplest way to take account of the terms of chemical production is to write them, as we did for a perfectly stirred reactor (see section 2.2), where we considered averaged values for the concentrations and temperature, like in equation [A3.45]. We then have:

$$\tilde{\dot{c}}=-B\bar{\rho}(1-\tilde{c})\exp\left[-\frac{T_a}{T_f+(T_b-T_f)\tilde{c}}\right] \qquad [3.54]$$

This model is pertinent only for small turbulent Damköhler numbers, i.e. if $\tau_t/\tau_{ch} << 1$.

3.3.4.7. *Production terms and pdf*

The chemical production terms are not linear. We cannot treat them in the same way as the turbulence transfer coefficients, and they cannot be interpreted simply by theories such as Prandtl's. The probability density function (pdf) is defined for a phase space which includes the parameters of state of the flow *T*, *v* and Y_j, as well as the spatial and temporal coordinates [AND 80]. If ζ is a vector which has these parameters as its components, we can define the pdf $P(\zeta,\mathbf{x},t)$ such that $\int_V P(\zeta,\mathbf{x},t)\,d\zeta=1$. The average value corresponding to the function $\varphi(\zeta,\mathbf{x},t)$ becomes $\bar{\varphi}(\mathbf{x},t)=\int_V \varphi P\,d\zeta$.

If we know the pdf, we can use it to deduce, e.g., the average species production rate $\bar{\dot{W}}(\mathbf{x},t)=\int_V \dot{W}(\zeta)P(\zeta,\mathbf{x},t)\,d\zeta$. The vector ζ will have the concentrations and the temperature as components, but for turbulent diffusion fluxes, it will be necessary to use a pdf dependent on the concentrations and the velocity components.

The pdf obeys a balance equation in the phase space, which is similar to the population balance in Chapter 2.

To obtain the closure conditions, which are necessary to solve the equations for turbulent flows, we need to have the balance equation for the pdf or to have provided an appropriate form of pdf in advance (this is known as the presumed pdf method).

Of the various pdf balance equations that we can choose, we shall select the simplest, where the phase space contains only one concentration coordinate and is independent of $\mathbf{x}$:

$$\frac{\partial P(C,t)}{\partial t}+\alpha\left[P(C,t)-P_0(C,t)\right]+\frac{\partial}{\partial C}\left[\dot{W}(C)P(C,t)\right]=$$
$$2\beta\left[\int P(C',t)P(C'',t)\delta\left(\frac{C'+C''}{2}-C\right)dC'dC''-P(C,t)\right] \qquad [3.55]$$

The first term on the left-hand side of this equation is the unsteady variation in P, the second corresponds to convection; α is the inverse of the average residence time and P_0 is the pdf at the reactor opening. The third term corresponds to chemical production – it is a term of flux in the phase space. Finally, the right-hand side of the equation takes account of the interactions between the fluid particles, assuming that two particles of concentration C' and C'' combine to form a third particle of concentration $(C' + C'') / 2$, with a frequency of coalescence/re-dispersion β (supposed to be constant). This equation, where we suppose that ρ = *const.*, does not take account of the inhomogeneities.

Other models are more sophisticated. Some such models use "two-point" pdf's. In these cases, we need to find good closure conditions for the pdf balance, and then solve it.

The pdf can be measured. In case of combustion involving a mixture of burned and fresh gases, the pdf of temperature will generally be bimodal, but its shape can evolve over the course of the reaction. A complete case is discussed using pdf's in [PET 82].

3.3.5. *LESs in combustion*

Here, we give a supplement to the part of Appendix 3 relating to large eddy simulations, introducing the balance equations and certain closure laws specific to combustion.[13]

3.3.5.1. *Balance equations filtered for combustion*

The equations in [A3.65] are not valid for combustion. We have to consider compressible flows and use the Favre average, and introduce the concentration variables, the rates of chemical production and the heat released. The resulting equations are similar in form to equation [A3.56] for mass and momentum and to equation [3.41] for the scalar quantities. We shall write them (see Poinsot and Veynante [POI 05]) in the following form, where $h_{s,j} = \int_{T^0}^{T} c_{p,j} dT$ and $h_s = h - \sum_j Y_j \left(q_f^0\right)_j$, the sensible heat corresponding to β_T in section 2.5.2 [PRU 12] in the case of a single chemical reaction: $h_s = \sum_j Y_j \int_{T^0}^{T} c_{p,j} dT = \beta_T \Delta H$.

We make the approximation $\overline{\mathbf{v}\cdot\nabla p} \cong \tilde{\mathbf{v}}\cdot\nabla\bar{p}$:

$$\begin{cases} \partial\bar{\rho}/\partial t + \nabla\cdot(\bar{\rho}\,\tilde{\mathbf{v}}) = 0 \\ \partial(\bar{\rho}\,\tilde{\mathbf{v}})/\partial t + \nabla\cdot(\bar{\rho}\,\tilde{\mathbf{v}}\otimes\tilde{\mathbf{v}}) + \nabla\bar{p} = -\nabla\cdot\left[\bar{\mathbf{\Pi}} + \bar{\rho}\left(\widetilde{\mathbf{v}\otimes\mathbf{v}} - \tilde{\mathbf{v}}\otimes\tilde{\mathbf{v}}\right)\right] \\ \bar{\rho}\,\partial\tilde{Y}_j/\partial t + \bar{\rho}\,\tilde{\mathbf{v}}\cdot\nabla\tilde{Y}_j = \nabla\cdot\overline{\rho D_j \nabla Y_j} - \nabla\cdot\left[\bar{\rho}\left(\widetilde{\mathbf{v}Y_j} - \tilde{\mathbf{v}}\tilde{Y}_j\right)\right] + \overline{\dot{W}_j} \\ \bar{\rho}\,\partial\tilde{h}_s/\partial t + \bar{\rho}\,\tilde{\mathbf{v}}\cdot\nabla\tilde{h}_s = \partial\bar{p}/\partial t + \overline{\mathbf{v}\cdot\nabla p} + \nabla\cdot\overline{\lambda\nabla T} - \nabla\cdot\left[\bar{\rho}\left(\widetilde{\mathbf{v}h_s} - \tilde{\mathbf{v}}\tilde{h}_s\right)\right] \\ -\overline{\mathbf{\Pi}:\nabla\otimes\mathbf{v}} - \nabla\cdot\overline{\rho\sum_j \mathbf{V}_j Y_j h_{s,j}} + \overline{\dot{W}_T} \end{cases} \qquad [3.56]$$

The filtered fluxes of molecular diffusion are neglected or modeled with a simple gradient hypothesis such as $\overline{\rho D_j \nabla Y_j} = \bar{\rho} D_j \nabla\tilde{Y}_j$, $\overline{\lambda\nabla T} = \bar{\lambda}\nabla\tilde{T}$.

13 These considerations, which are part of Chapter 2, "Turbulent flow concepts", in [PRU 10], are presented separately in this book, Chapter 3 of which is devoted specifically to combustion.

LESs are used in different cases of flows – compressible or incompressible, more or less confined and homogeneous [DEA 70; DUC 96; SUL 03]. We are able to solve certain problems of combustion using LESs (see the work of the CERFACS team, for example [ROU 05; SEL 04]). Poinsot and Veynante [POI 05] give simple closures for the chemical production rates, which are supposed to apply both to premixed and diffusion flames.

3.3.5.2. *Closure equations filtered for combustion*

The simplest closure relation for non-resolved LES fluxes is a relation of gradient: $\widetilde{\mathbf{v}Y_j} - \tilde{\mathbf{v}}\tilde{Y}_j = -\left(\nu_t / Sc_j\right)\nabla\tilde{Y}_j$.

For the LES-filtered reaction rate, we describe two simple models which are supposed to apply to all types of flames. There are also other specific models; these shall not be presented here, as they are described in [POI 05].

3.3.5.2.1. Arrhenius model for filtered quantities

This model corresponds to the Arrhenius model in section 3.3.4, but this time for the filtered values under discussion here. We work with the assumption of a perfect mixing on the sub-grid scale and discount the fluctuations on this scale. It is supposed that $\tau_t << \tau_c$, where τ_t is the sub-grid turbulent timescale and τ_c the chemical timescale. Thus, for a second-order chemical reaction, we have $\overline{\dot{W}}_j = \nu'_j B\bar{\rho}^2\bar{T}^s\bar{Y}_F\bar{Y}_O e^{T_a/T}$, where B and s are the constant coefficients in Arrhenius' law (Chapter 2 of [PRU 12]) and $T_a = E_a/R$ is the reaction activation temperature.

If the fuel and the oxidizer are not perfectly mixed at the sub-grid level, we can write the following formula with a segregation factor $\left(\widetilde{Y_F Y_O} - \tilde{Y}_F\tilde{Y}_O\right)/\tilde{Y}_F\tilde{Y}_O$:

$$\overline{\dot{W}}_j = \nu'_j B\bar{\rho}^2\bar{T}^s\bar{Y}_F\bar{Y}_O e^{T_a/T}\left(1 + \frac{\widetilde{Y_F Y_O} - \tilde{Y}_F\tilde{Y}_O}{\tilde{Y}_F\tilde{Y}_O}\right) \qquad [3.57]$$

3.3.5.2.2. Modeling of reaction rates with scale-similarity hypotheses

The scale-similarity hypotheses have been extended to the modeling of the reaction rates [GER 97].

3.3.6. *Triple flames*

In a mixture presenting inhomogeneities in concentration, we may witness the development of triple flames (also known as tribrachial flames).

Remember firstly that the normal speed of steady laminar combustion in an adiabatic premixed plane flame is s_L^0. Its value is obtained in a stoichiometric mixture, in the presence of a dominant reaction between a fuel F and an oxidizer O to give the combustion products P, with an equivalence ratio equal to 1. This speed s_L^0 is in fact a maximum of the speed of combustion s_L that can be obtained by varying the equivalence ratio of the mixture. Thus, when we increase the scalar concentration variable Z from its minimum to its maximum so that combustion can take place, the speed s_L experiences a maximum s_L^0 at $Z = Z_{st}$. The same is true in a turbulent regime with the speed of turbulent combustion s_T, which reaches a maximum s_T^0 greater than s_L^0 (see [ECH 96; PET 00]).

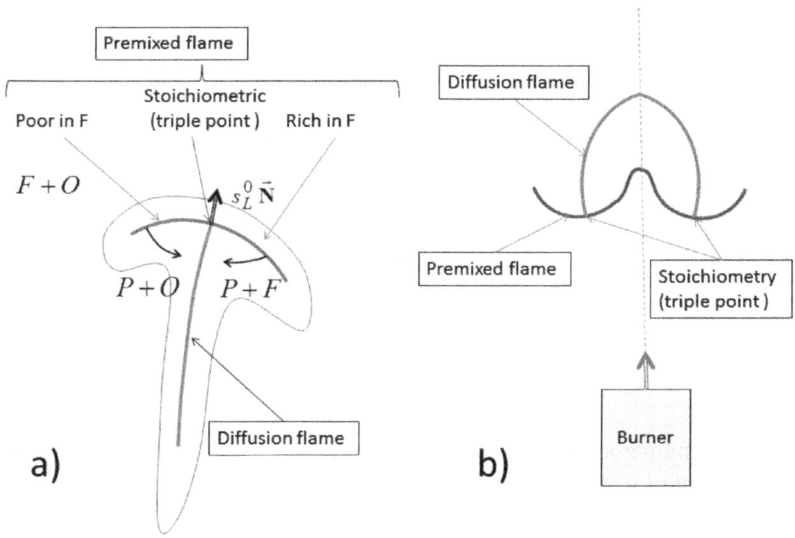

Figure 3.16. *Triple flames: a) in a non-homogeneous mixture; b) lifted flame*

If a premixed flame arises in a non-homogeneous mixture at a point with stoichiometric concentration, it expands with a poor branch and a rich branch, and becomes curved on each side of the summit, given that its normal speed of propagation is slower the poorer or richer the mixture. That is to say, the propagation speed is slower the further we get from the summit (or from the line of stoichiometric concentration).

However, there is a difference, at the flame front, between the poor side where the oxidizer O mixed with the combustion products P is ejected, and the rich side where the fuel F mixed with the combustion products P is ejected (more precisely, for a hydrocarbon flame, on one side we find oxidizing intermediary products and on the other we find reducing ones). These components, which are likely to react in front of the premixed flame can give rise to a diffusion flame connected to the stoichiometric line of the premixed flame and, if that flame is thin, its position will be determined by $Z = Z_{st}$. This is the triple flame phenomenon as illustrated in Figure 3.16.

In certain circumstances, both parts, poor and rich, of the premixed flame disappear and we are left with only the stoichiometric summit. The triple flame has degenerated into a single "diffusion flame" part. This situation was first studied by Buckmaster [BUC 96], who cites the example of the leading edge of a flame which is spreading over a bed of solid or liquid fuel. These so-called *edge flames* have a leading edge corresponding to the triple point, where the speed of combustion is that of the stoichiometric premixed flame. They also have a trailing edge where extinction takes place, and their form is thicker at their base. The circumstances in which they appear are linked to the stretch rate of the flow, which must be sufficient for the thickness of diffusion on both sides to be similar to the thickness of the zone of preheating of the premixed flame [NAY 02].

Peters [PET 00] gave a complete review of the problem. Besides the general mechanism, we can see the advantage for specialists in combustion to become more familiar with triple flames. We can cite the remark made by Domingo and Vervisch [DOM 96] about self-ignition – that "a triple flame is capable of surviving significant interactions with vortices, adjusting its structure to a transitory environment".

Since then, numerous studies have been done and are still being done today on triple flames. Let us cite some of the most recent.

Nayagam and Williams [NAY 02] look at the effects of the Lewis number on the propagation of *edge flames*, which are an extreme case of triple flames. A simplified model is developed. Figure 3.17 shows the diagram for this model.

The configuration is not dissimilar to that of the boundary layer with chemical reactions (the Emmons problem) studied in microgravity by the team at the LCD (Combustion and Detonics Lab) in Poitiers, France [PRU 10].

Using an asymptotic analysis such as that performed by Clavin and Joulin, the authors obtain the expression of the edge propagation speed as a function of the Damköhler number.

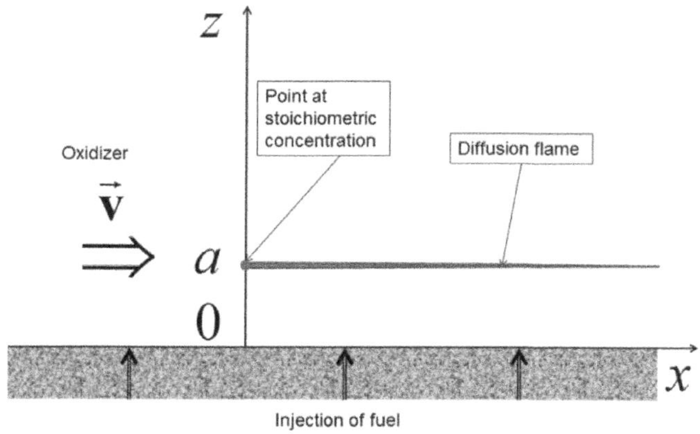

Figure 3.17. *Schematic illustration of an edge flame (redrawn from [NAY 02])*

Figure 3.18 gives the velocity of propagation V as a function of the Damköhler number δ using the following relations:

$$\begin{cases} V+\sqrt{V^2+4}=2\varepsilon_\infty\sqrt{2\Delta_\infty}\exp\left[\left(\ell V/4\sqrt{V^2+4}\right)\left(1+V/\sqrt{V^2+4}\right)\right] \\ \delta=\varepsilon_\infty\sqrt{2\Delta_\infty}=\left(RT_\infty^{\ 2}/E\left(q/c_p\right)\right)\sqrt{2\rho c_p at\, A/\lambda}\exp\left(-E_a/2RT_\infty\right) \end{cases} \quad [3.58]$$

In these relations, $V=v\rho c_p\sqrt{at}/\lambda$ with a being the thickness of the diffusion layer, $t=\varepsilon a$ the thickness of the reactive layer with $a \ll 1$, v the component along x in the velocity of the flow (u along z), $\Delta=2\rho c_p at\, Ae^{-E_a/RT_f}/\lambda$ the Damköhler number, the ratio of diffusion time to chemical time, with T_f being the temperature of the premixed flame, E_a is the reaction activation energy, A its Arrhenius coefficient and q the heat released by that reaction. The results show that with large Damköhler numbers, an increased Lewis number increases the combustion rate. With small Damköhler numbers, the opposite is true, so an increased Lewis number decreases the combustion rate. For a given Damköhler number, there can be three steady solutions, with the middle one certainly being unstable.

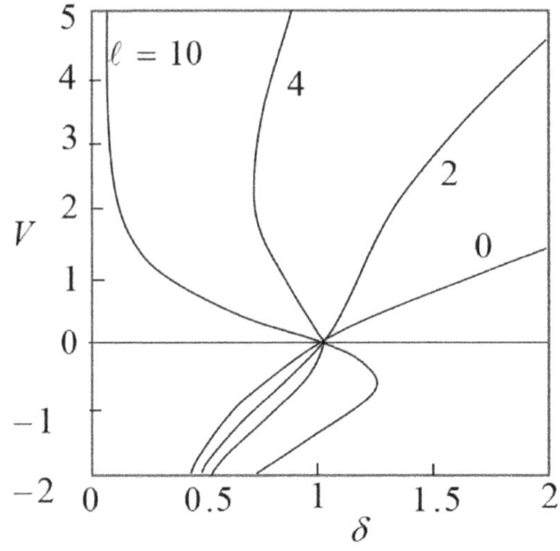

Figure 3.18. *Flame edge velocity as a function of the Damköhler number for different values of ℓ (redrawn from [NAY 02])*

Ko *et al.* focused on triple flames in counter-current flows [KO 02], which we mentioned in sections 3.1.3.2 and 3.2.2.2.

Lyons [LYO 07] conducted experiments to gain a fuller understanding of the mechanisms of stabilization of flames spinning in a turbulent jet. These mechanisms, which are sometimes highly complex, involve triple flames.

APPENDICES

Appendix 1

Dimensionless Numbers, Similarity

As an accompaniment to this volume devoted to typical examples of flows which can include chemical reactions, it is helpful to introduce certain analytical tools which are very useful in approaching any physical problem. These tools together make up dimensional analysis [PRU 10], whose field of application extends far beyond laminar flows.

The dimensionless numbers are products of quantities, raised to certain powers which are determined analytically. These quantities are parameters concerning the problem to be studied: transfer coefficients, dimensions of a reactor, average thermal gradient, characteristic time of a phenomenon, etc. They must be expressed in a coherent system of units.

These numbers are very useful in performing a preliminary evaluation of the prevailing phenomena in a flow. Yet the advantage to their use does not stop here. Alterations in stability are often characterized by curves or surfaces of a space defined using such numbers. Sometimes it is possible to study certain phenomena even without complete solving of the equations which govern them, and to draw very useful conclusions on the basis of the dimensionless numbers relating to the experimental observations.

In this appendix, we give a very succinct presentation of dimensional analysis. We shall see that in addition to the numbers cited, the concepts which stem from this also play a part in determining the solutions to the equations for a given problem, revealing groups of variables that are of interest, and also in the implementation of similarity experiments. A table of commonly-used dimensionless numbers is given at the end of this appendix.

A1.1. Fundamentals of dimensional analysis: Π_i groups

A1.1.1. *Basic considerations*

A given problem is such that the solution furnishes relations between the characteristic quantities $F_1, F_2, \ldots F_n$ of the medium in question. These quantities are measured on the basis of quantities which all belong to a coherent system of units.

The basic idea in dimensional analysis is founded on the *conviction that the physical relations are independent of the units chosen* to evaluate these quantities.

Consider

$$f(F_1, F_2, \ldots F_n) = 0 \qquad [A1.1]$$

a final relation between *n* quantities of a given problem. We know that the form of this relation will not depend on the system of units used to measure these quantities.

If $F_i = x_i e_i$, in a given system of units e_i, then in a system of units e'_i, we would have

$$x'_i e'_i = x_i e_i \qquad [A1.2]$$

From a dimensional point of view, all quantities are expressed by way of the primary quantities, of which there are three: length *L*, mass *M* and time *T*.

Therefore, symbolically, we write

$$[F_i] = \left[L^{\alpha_i} M^{\beta_i} T^{\gamma_i}\right] \qquad [A1.3]$$

For instance, the pressure is expressed as $[p] = \left[L^{-1} M T^{-2}\right]$.[1]

We can envisage basic quantities other than the primary quantities. If

$$[X] = \left[L^{a_{11}} M^{a_{12}} T^{a_{13}}\right],\ [Y] = \left[L^{a_{21}} M^{a_{22}} T^{a_{23}}\right],\ [Z] = \left[L^{a_{31}} M^{a_{32}} T^{a_{33}}\right] \qquad [A1.4]$$

1 Certain values are defined on the basis of the primary values and universal constants. In this case we overlook the dimensions of these constants. For instance, the temperature θ simply gives us $[\theta] = \left[L^2 T^{-2}\right]$.

the dimensional equation of the quantities F_i will be, indifferently

$$[F_i]=\left[L^{\alpha_i}M^{\beta_i}T^{\gamma_i}\right] \text{ or } [F_i]=\left[X^{x_i}Y^{y_i}Z^{z_i}\right] \quad \text{[A1.5]}$$

on condition that the matrix of the a_{kl} of the transformation for converting from one to the other is regular. Indeed, by identifying the two expressions of F_i, we obtain

$$\begin{cases} a_{11}\,x_i + a_{21}\,y_i + a_{31}\,z_i = \alpha_i \\ a_{12}\,x_i + a_{22}\,y_i + a_{32}\,z_i = \beta_i \\ a_{13}\,x_i + a_{23}\,y_i + a_{33}\,z_i = \gamma_i \end{cases} \quad \text{[A1.6]}$$

which enables us to calculate the coefficients x_i, y_i, z_i.

Suppose that the table of coefficients $\alpha_i, \beta_i, \gamma_i$, exponents of the *L, M* and *T,* is of $p(p \le 3)$ rank. In this case, there are *(n-p)* dimensionless products, constituted of the *p* fundamental values taken from the *n* values and classified in such a way that

$$X = F_1, Y = F_2, \ldots \quad \text{[A1.7]}$$

Thus, the products are of the form

$$\Pi_i = \frac{F_i}{F_1^{x_i} F_2^{y_i} \ldots}, i = p+1, \ldots n \quad \text{[A1.8]}$$

A1.1.2. *Vaschy–Buckingham theorem (1890) or Π theorem*

A final relation $f(F_1, F_2, \ldots F_n) = 0$ between *n* values can be placed in the form of an equation between (*n* – *p*) groups, *p* being the maximum number of dimensionally independent quantities which intervene in the equation in question. Therefore, we write

$$\varphi(\Pi_{p+1}, \ldots \Pi_n) = 0 \quad \text{[A1.9]}$$

or indeed

$$\Pi_{p+1} = \psi\left(\Pi_{p+2},\ldots\Pi_n\right) \qquad [A1.10]$$

Note at this point that dimensional analysis does not give us the functions φ and ψ. Only experimentation can give us the form of these functions (or complete mathematical analysis of the problem).

A1.1.3. *Practical advantage to dimensional analysis*

We find the following advantages [DAR 96]:

– reduction of the number of variables (sometimes by way of some additional hypotheses);

– search for self-similar solutions;

– presentation of results;

– search for solutions by experimental or numerical means.

The step-by-step approach is outlined below:

1. Include in f all the quantities which might possibly be found (independent variables, physical constants) and then eliminate those whose effect is presumed to be negligible.

2. For basic quantities X, Y, choose those whose intervention is certain, which it is difficult to keep constant.

3. Form the Π groups.

4. Formulate the law $\varphi\left(\Pi_{p+1},\ldots\Pi_n\right)=0$ (or the ψ law). This law can sometimes be expressed in the form $\Pi_{p+1}{}^{\alpha}\,\Pi_{p+2}{}^{\beta}\ldots = \text{const.}$

A1.1.4. *Example of application: head loss in a cylindrical pipe*

The roughness of the wall is characterized by an average thickness e (Figure A1.1).

1. The dimensions of the active quantities are given in Table A1.1. This is the first step.

F_i	D	Δl	e	ρ	V	Δp	μ
L	1	1	1	-3	1	-1	-1
M	0	0	0	1	0	1	1
T	0	0	0	0	-1	-2	-1

Table A1.1. *Dimensions of the active values*

Let us now follow the other steps indicated above.

2. Choice of basic quantities $\{X, Y, Z) = \{D, \rho, V\}$ (the base-change matrix is regular). Thus, we have (Table A1.1):

$$[\Delta l]=[e]=[D]=[L],[V]=\left[LT^{-1}\right],[\Delta p]=\left[L^{-1}MT^{-2}\right],$$

$[\mu]=\left[L^{-1}MT^{-1}\right]=\left[X^{x_\mu}Y^{y_\mu}Z^{z_\mu}\right]$. We find $[\Delta p]=\left[\rho V^2\right]$, $[\mu]=[\rho DV]$

3. The Π groups are therefore $\Pi_{\Delta l}=\Delta l/D$, $\Pi_e=e/D=\varepsilon$ (relative roughness), $\Pi_{\Delta p}=\Delta p/\rho V^2$, $\Pi_\mu=\mu/\rho VD=1/\mathrm{Re}$ (the inverse of the Reynolds number).

4. Vaschy–Buckingham theorem:

$$\Delta p/\rho V^2=\psi_1(\Delta l/D,\varepsilon,\mathrm{Re}) \qquad [A1.11]$$

Experimental result:

$$\Delta p/\rho V^2=(\Delta l/D)\psi_2(\varepsilon,\mathrm{Re}) \qquad [A1.12]$$

Significant parameter: $\Lambda=2D\Delta p/\rho V^2\Delta l$, head loss coefficient.

Thus, $\Lambda=\psi(\varepsilon,\mathrm{Re})$. Experimentation confirms this result in both laminar and turbulent regimes.

In particular, in the laminar regime:

$$\Lambda=64/\mathrm{Re} \qquad [A1.13]$$

as shown by the calculation performed in Chapter 1 of [PRU 12] (section 1.8.6.2).

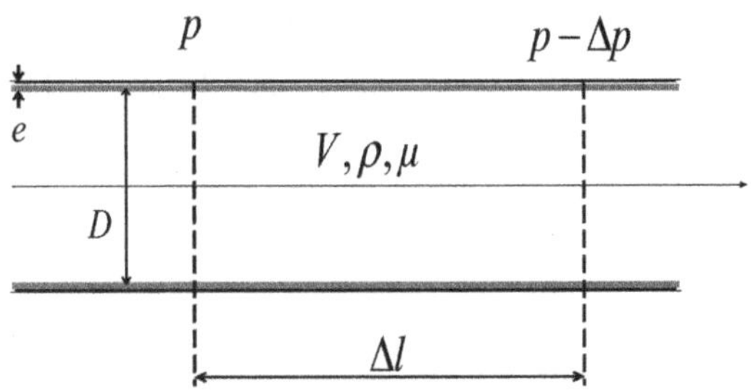

Figure A1.1. *Head loss in a cylindrical pipe*

A1.2. Similarity

A1.2.1. *Definition*

Our aim here is to represent a real phenomenon, generally involving a prototype (hydraulic construction, aeronautics, machines) as a similar phenomenon, using a reduced-scale model.

There is a certain number of so-called "similarity" rules that need to be respected so that the solution to the problem on the model is transposable to the real-world problem.

The principles of dimensional analysis are applicable, because changing the units used to measure the values F_i is equivalent to considering two systems S and S' for the same physical problem, with the homologous values measuring x_i and x'_i in the same unit.

When all the similarity conditions are met, we say that there is perfect similarity. This is often difficult to achieve. With each value, we associate a scaling factor λ_i. If the similarity is perfect, we must have both of the following conditions simultaneously:

$$f\left(F_1,F_2,\ldots F_n\right)=0 \text{ and } f\left(\lambda_1 F_1,\lambda_2 F_2,\ldots \lambda_n F_n\right)=0 \qquad \text{[A1.14]}$$

The Π groups must be such that

$$\Pi_i=\frac{F_i}{F_1^{x_i}F_2^{y_i}\ldots}=\frac{\lambda_i F_i}{\lambda_1^{x_i}F_1^{x_i}\lambda_2^{y_i}F_2^{y_i}\ldots},\ i=p+1,\ldots n \qquad \text{[A1.15]}$$

if we wish to verify the Vaschy–Buckingham theory, ensuring the invariance of the relations $\varphi(\Pi_{p+1},\ldots\Pi_n)=0$. In this case, we would have:

$$\frac{\lambda_i}{\lambda_1^{x_i}\lambda_2^{y_i}\ldots}=1,\ i=p+1,\ldots n \qquad [A1.16]$$

In addition, we need to ensure that the usual simplifications are still valid for the model (e.g. the phenomenon of viscosity may be negligible in one case but not in another).

A1.2.2. *Application: condition of similarity in a soft balloon placed in a current of air with a given velocity*

We propose to create a model of a large balloon, so as to obtain the same tension on the skin (Figure A1.2).

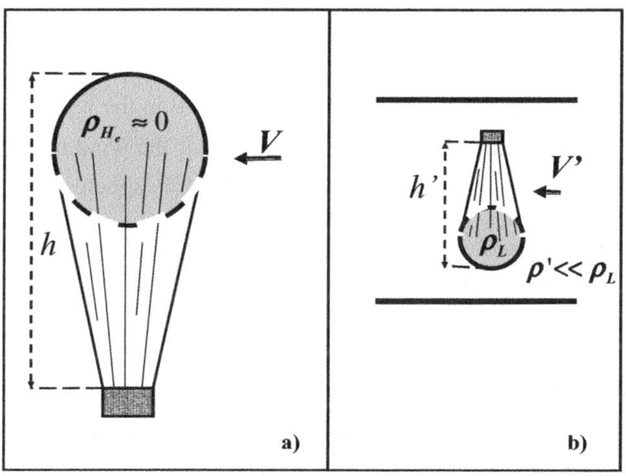

Figure A1.2. *a) Stratospheric balloon; b) model in a wind tunnel*

Many different phenomena are involved. The first is the tension due to hydrostatic pressure $\Delta p \propto \rho g h$. We have the surface tension $\sigma \propto R\Delta p$ (*R* being the radius of the balloon).

We therefore need to have $R\,\Delta p = R'\Delta p'$ (R' being the radius of the model) so that the tension is the same on the skin of the balloon and that of the model, or indeed $R\,\rho\,h = R'\rho' h'$ so $\rho'/\rho = Rh/R'h' \cong (R/R')^2 = \lambda^2$.

Because the model has to be smaller than the real balloon, ρ' has to be greater than ρ.

Yet the problem then arises of how to ensure that the model balloon retains its shape in a wind tunnel. We can solve this problem by filling it with a liquid with density ρ_L . The above results remain valid, with the model being hung upside down (see Figure A1.2b): $\rho_L/\rho \cong (R/R')^2 = \lambda^2$. With water, $\lambda^2 = 1000$, so in approximate terms, $R/R' = \lambda \cong 30$.

Furthermore, if ρ' is the density of the air around the model, the aerodynamic pressure, in an incompressible flow (Mach number $M < 0.25$), is $\Delta p \propto \rho V^2$. Because the tensions caused on the skin of the balloon need to be equal, we have $\rho V^2 R = \rho' V'^2 R'$, and it follows that $R/R' = \rho' V'^2/\rho V^2$.

If the air has the same density in the stratosphere and in the wind tunnel, we find $V'/V = \sqrt{\lambda}$. With water in the model, this implies that the wind speed in the wind tunnel needs to be five to six times greater than in the stratosphere.

Now consider the viscous stresses; these bring into play the Reynolds numbers Re and Re' such that $\dfrac{\text{Re}'}{\text{Re}} = \dfrac{\rho' V' R'}{\mu'} \dfrac{\mu}{\rho V R}$, where $R/R' = 1/\lambda$.

If the temperature and pressure are unchanged, ρ and μ are the same in the stratosphere and in the wind tunnel, and it follows that $\text{Re}'/\text{Re} = V'R'/VR = 1/\sqrt{\lambda}$.

In this case, we do not have similarity of the Reynolds numbers. In order to achieve this similarity we can increase the pressure at constant temperature (so that $\rho \propto p$). We explicitly write

$$\frac{\text{Re}'}{\text{Re}} = 1 = \frac{\rho' V' R'}{\rho V R} = \sqrt{\frac{p'}{p}}\sqrt{\frac{\rho' V'^2}{\rho V^2}}\sqrt{\frac{R'}{R}} = \sqrt{\frac{p'}{p}}\frac{1}{\sqrt{\lambda}}$$

Therefore, we must satisfy $p'/p = \lambda$ (with the example of water, $p' \approx 30p$).

We can also decrease the temperature (using a cryotechnic wind tunnel):

$$\frac{\text{Re}'}{\text{Re}} = 1 = \sqrt{\frac{\rho'}{\rho}}\sqrt{\frac{\rho' V'^2}{\rho V^2}}\sqrt{\frac{R'}{R}}\sqrt{\frac{T}{T'}}$$

because $\rho \propto p/T$ and $\mu \propto \sqrt{T}$. If we impose $p' = p$, we find that $T'/T = 1/\sqrt{\lambda}$.

In water, which needs to be heated so as not to transform into ice, we find approximately $T' \approx T/5$.

A1.3. Analytical searching for solutions to a heat transfer problem (self-similar solution)

The heat transfer equation in a continuous medium at rest was established in Chapter 1 of [PRU 12]. The energy balance equation is written as

$$\rho \frac{de}{dt} + \vec{\nabla} \cdot \vec{\mathbf{q}} = 0 \qquad \text{[A1.17]}$$

We suppose the density to be constant, the internal energy to be proportional to the temperature $e = cT$ with a constant calorific capacity and the heat flux to verify a Fourier law $\vec{\mathbf{q}} = -\lambda \vec{\nabla} T$ with constant heat conductivity. We finally obtain, for an evolution in flat layers:

$$\partial T/\partial t = \kappa \partial^2 T/\partial x^2 \qquad \text{[A1.18]}$$

with constant heat diffusivity $\kappa = \lambda/\rho c$ (see section 1.8.3.2).

The solution to a given problem is of the form:

$$T = f(x,t,\kappa) \qquad \text{[A1.19]}$$

We look for solutions which, for coefficients λ, μ, ν to be determined, give us

$$f(x,t,\kappa) = f(\lambda x, \mu t, \nu \kappa),\ \forall x,t,\kappa \qquad \text{[A1.20]}$$

In other words, the solution $f(x,t,\kappa)$ must be invariant in a change of spatial scale, time and heat diffusivity.

We know that $f(x,t,\kappa)$ takes the form of a dimensionless group. Indeed, we can easily establish that $\Pi_\kappa = \kappa t/x^2$; hence $T = \varphi(x^2/\kappa t)$. Because $x^2/\kappa t = (\lambda^2/\mu v)x^2/\kappa t$, we obtain the condition $\lambda^2 = \mu v$.

Therefore, we look for solutions in the above form. These indeed exist in certain circumstances, i.e. if the initial data and boundary data facilitate it.

Let us pose the specific problem of heat transmission in a semi-infinite medium (Figure A2.1). This problem was treated in the exercise in section 1.8.3.2 [PRU 12]. A medium occupying the space $x \geq 0$ whose temperature is uniform T_1, is brought into contact, at time $t = 0$, with a thermostat at temperature T_0.

If we set $\eta = x/\sqrt{4\kappa t}$, as we are encouraged to do by the expression of the group Π_κ above (a choice which was made *a priori* in section 1.8.3.2), we obtain $\frac{d^2T}{d\eta^2} + 2\eta\frac{dT}{d\eta} = 0$.

The solution is $T = T_0 + (T_1 - T_0)\operatorname{erf}\eta$, where $\operatorname{erf}\eta = \frac{2}{\sqrt{\pi}}\int_0^\eta e^{-s^2} ds$. Figure 1.36 in [PRU 12] shows the evolution over time of the reduced temperature profile.

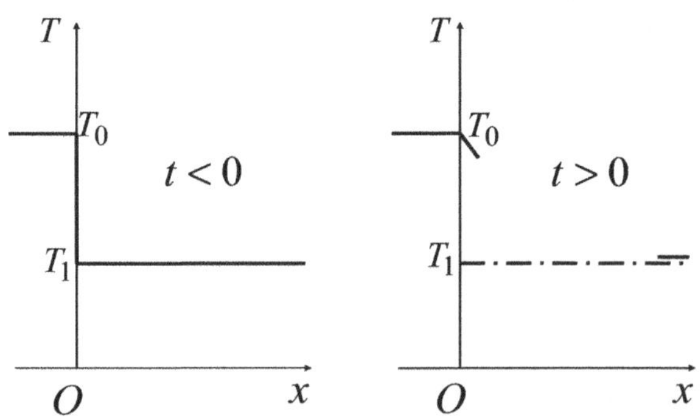

Figure A1.3. *Boundary conditions and initial conditions for the "thermal shock" problem discussed in section 1.8.3.2 [PRU 12]*

A1.4. Some dimensionless numbers

Similarities in fluid mechanics respect the ratios between various types of force which are involved in motion. As in most flows, it is inertial forces, which are necessary to alter the motion of fluid particles, which are most prevalent. These are then compared to other types of forces: gravity, viscosity, surface tension and compressibility. Thus, we find the Froude, Reynolds, Weber and Mach numbers.

Thus, the value of the *Froude number* gives us an indication of the order of magnitude of the relative values of the inertial forces in relation to gravitational forces. The Froude similarity respects the ratio between the forces of inertia and the forces due to gravity; it is imposed when the position and motion of the fluid particles can be influenced by these two types of force, and particularly with free surface flow.

The *Reynolds number* characterizes the relative significance of the forces of inertia and the forces of viscosity. The Reynolds similarity respects the ratio between these forces.

It is important to note the general result which, for any problem involving both viscosity and gravity forces, similarity-based study – which is often the only possible means – simultaneously reveals the Froude and Reynolds numbers. Preservation of both of these numbers renders perfect similarity an impossibility. In this case, we are led to seek *approximate similarity* by preserving only one of these two numbers (obviously the one which seems more important). In particular, this is what happens with problems studied in towing tanks.

The *Grashof number* is presented as the square of a particular Reynolds number associated with a velocity induced by flotation forces $V_G = \sqrt{Lg\Delta\rho/\rho}$.

The *Weber number* characterizes the relative values of the forces of inertia and surface tension. A Weber similarity can only be sought if there is a free surface where surface tension phenomena can manifest themselves.

The *Mach number* characterizes the relative value of the inertia and elastic forces of compression of the fluid (indeed, $c = \sqrt{(\partial p/\partial \rho)_s}$). The Mach similarity will be respected in problems where the phenomenon of compressibility plays a fundamental role, and each time there is wave propagation.

The *Rossby number* enables us to compare the effect of the inertial forces to the effect of the Coriolis forces.

Most dimensionless numbers can be defined as *ratios of characteristic times*. Such is the case with the Reynolds number, introduced above, which is also equal to the ratio between a time of diffusion by viscosity of the momentum L^2/ν and a momentum convection time L/V. The *Peclet number* is equal to the ratio between a heat diffusion time L^2/κ and a convection time L/V. In the same manner, we also define the *Prandtl number*, which is equal to the heat diffusion time / momentum diffusion time ratio; the *Schmidt number*, equal to the species diffusion time / momentum diffusion time ratio; the *Lewis number*, equal to the ratio of species diffusion time / heat diffusion time; the Strouhal number *Sr* (convection time / eddy emission characteristic time); the Mach number (acoustic time L/c / convection time). The first *Damköhler parameter Da* is equal to the ratio of a chemical time to a mechanical time. We can define the *Marangoni number Ma* as being the ratio of a heat diffusion time to a capillary time corresponding to the velocity induced by the spatial variation of the surface tension $V_\sigma = \Delta\sigma/\mu$. The *Ohnesorge number Oh* is defined as the ratio of a capillary time $\sqrt{\rho L^3/\sigma}$ to a viscous diffusion time L^2/ν.

Certain numbers are defined as products of other dimensionless numbers, and others as ratios of flux (such as the Nusselt number).

Table A1.2 gives a list of the dimensionless numbers commonly used in fluid mechanics.

Name of dimensionless number	**Expression of dimensionless number**	**Associated phenomenon**
Bond number	$Bo = gL^2\,\Delta\rho/\sigma$	Superposed fluids with surface tension
Crispation number	$Cr = \mu\kappa/\sigma^2\,\delta$ (section 3.2.3 in [PRU 12])	Heat diffusion, viscosity, surface tension
Damköhler number	$Da = \tau_{mech}/\tau_{ch}$	Chemical reaction in a flow
Froude number	$Fr = V^2/gL$	Inertia and gravity
Chemical/thermal Grashof numbers	$Gr_{ch} = gL^3\Delta\rho_{ch}/\rho\nu^2$ $Gr_{th} = gL^3\Delta\rho_{th}/\rho\nu^2$	Flotation and viscosity

Table A1.2. *Dimensionless numbers*

Hickman number	$Hi = \frac{\dot{m}^* \beta \delta^2 \mu_V}{\sigma^* \rho_L \kappa_L} \frac{d\dot{m}^*}{dT} \left(\frac{1}{\rho_V} - \frac{1}{\rho_L} \right)$	Differential vapor recoil
Lewis number	$Le = Sc/\Pr = \kappa/D$	Heat and mass diffusion
Mach number	$M = V/c$	Convection and elastic compression forces
Marangoni number	$Ma_{ch} = \sigma_C G_C \delta^2 / D\mu$ $Ma_{th} = \sigma_T G_T \delta^2 / \kappa\mu$	Gradient of surface tension, diffusion and viscosity
Nusselt number	$Nu = \alpha L/\kappa$	Heat diffusion and convection coefficient
Ohnesorge number	$Oh = \sqrt{\rho_L v^2 / L\sigma}$	Surface tension and viscosity
Peclet number	$Nu = VL/\kappa$	Convection and heat conduction
Prandtl number	$\Pr = \nu/\kappa$	Viscosity and heat diffusion
Chemical/thermal Rayleigh numbers	$Ra_{ch} = Gr_{ch}\, Sc = gL^3 \Delta\rho_{ch} / D\mu$ $Ra_{th} = Gr_{ch} \Pr = gL^3 \Delta\rho_{th} / \kappa\mu$	Flotation, viscosity and chemical/thermal diffusion
Reynolds number	$\mathrm{Re} = \rho VL/\mu$	Convection and viscosity
Richardson number	$Ri = (\rho_L - \rho_G)\, g\, \delta_G / \rho_G (V_L - V_G)^2$	Superposed fluids and mixture layers
Schmidt number	$Sc = \nu/D$	Viscosity and diffusion of species
Sherwood number (chemical Nusselt number)	$Sh = \alpha_C L/D$	Diffusion of species and convection coefficient
Surface viscosity number	$Vi = (\kappa_a + \varepsilon_a)/\mu D$	Surface viscosities and viscosity of the fluid
Weber number	$We = \rho V^2 L/\sigma$	Inertia and surface tension

Table A1.2.(continued) *Dimensionless numbers*

COMMENTS ON TABLE A1.2.– In this table, σ denotes surface tension; g acceleration due to gravity; L is a characteristic length; ρ is density; $\Delta\rho = \rho^* - \rho$; μ is viscosity; $\nu = \mu/\rho$ kinematic viscosity; κ heat diffusivity; τ_{mech} is a mechanical time; τ_{ch} is a chemical time; p is pressure; V velocity; $\dot{m}$ is a unitary mass evaporation rate; (*) relates to a reference state; $(_G)$ to a gas; $(_L)$ to a liquid; $(_V)$ to its vapor; δ is the thickness of a fluid layer; $(-\beta)$ is the temperature gradient in the thermal boundary layer with thickness δ; D is the coefficient of binary diffusion of the species in a mixture; c is the speed of sound; $\sigma_C = d\sigma/dC$, $\sigma_T = d\sigma/dT$, G_C and G_T are respectively the gradients of concentration and temperature; α is a convective heat transfer coefficient which becomes α_C for chemical species; $(_{ch})$ refers to a chemical effect and $(_{th})$ to a thermal effect; and finally κ_a and ε_a are coefficients of surface viscosity.

Appendix 2

Thermodynamic Functions

A2.1. General points

The specific thermodynamic functions of each of the species which are involved in the definition of the free energy, of the entropy, of the enthalpy, the internal energy and the thermodynamic potential of the mixture are determined by way of statistical mechanics and by experimental measurements [HER 51; CAS 70; PEN 57; JAN 69; BAR 73]. In the case of Boltzmann's statistics, these values are evaluated on the basis of the molecular partition function Q defined by $Q = \sum_{l} C_l \exp(-\varepsilon_l / kT)$ where, for each state of energy of the group l of level ε_l, C_l defines the number of states of that group and k is Boltzmann's constant.

If N is the number of molecules in a volume $\mathcal{V}$, we obtain, for instance,

$$\begin{aligned} &G = -NkT \ln(Q/N), F = G - pV = -NkT[\ln(Q/N) + 1] \\ &S = -(\partial G/\partial T)_{p,N} = kN[\ln(Q/N) + T(\partial \ln Q/\partial T)] \end{aligned} \quad [A2.1]$$

or indeed:

$$\begin{aligned} &S = -(\partial F/\partial T)_{V,N} = kN\left[\ln(Q/N) + 1 + T(\partial \ln Q/\partial T)_{V,N}\right], \\ &H = G - T(\partial G/\partial T)_{p,N} = kNT^2 (\partial \ln Q/\partial T)_{p,N}, E = kNT^2 (\partial \ln Q/\partial T)_{V,N}, \\ &C_p = (\partial H/\partial T)_{p,N} = -T(\partial^2 G/\partial T^2)_{p,N}, C_v = (\partial E/\partial T)_{V,N} = -T(\partial^2 F/\partial T^2)_{V,N} \end{aligned}$$

[A2.2]

This partition function takes account of all the specific motions of the atom or the molecule, which we classify as either translational motion or internal motions. The total energy of the atom or molecule is the sum of the translational energy and the energy of the internal modes. The partition function relating to all energy levels is written as $Q = Q_t Q_{\text{int}}$.

A2.2. Translational motion

The translational motion independent of the structure of the element or the molecular edifice is therefore common to all the species present.

The partition function due to translational motion is obtained by considering, in a box with dimensions a_1, a_2, a_3 and volume $V = a_1 a_2 a_3$, the possible states of energy for a particle with mass M moving within the box:

$$Q_t = \sum_{n_1, n_2, n_3} \exp\left[-\left(\frac{n_1^2}{a_1^2} + \frac{n_2^2}{a_2^2} + \frac{n_3^2}{a_3^2}\right)\frac{h^2}{8MkT}\right]$$

the numbers n_1, n_2, n_3 taking the values of the series of positive whole numbers from 1 to infinity and h being Planck's constant.

The sum of this triple product is calculated approximately by setting

$$\sum_{n_1=1}^{\infty} \exp\left(-\frac{n_1^2}{a_1^2}\frac{h^2}{8MkT}\right) \cong \int_0^{\infty} \exp\left(-\frac{n_1^2}{a_1^2}\frac{h^2}{8MkT}\right) dn_1 = a_1 \sqrt{\frac{2\pi MkT}{h^2}}$$

The translational partition function is therefore of the form $Q_t = V\left(\sqrt{2\pi M kT/h^2}\right)^3$.

By working out the molar mass $\mathcal{M} = \mathcal{N}_0 M$ ($\mathcal{N}_0$ Avogadro's number) and the pressure p given by the equation of state $pV = \mathcal{N}_0 kT$, we get

$$\frac{Q_t}{\mathcal{N}} = \frac{V}{\mathcal{N}}\left(\sqrt{\frac{2\pi \mathcal{M} kT}{\mathcal{N}_0 h^2}}\right)^3 = \frac{kT}{p}\left(\sqrt{\frac{2\pi \mathcal{M} kT}{\mathcal{N}_0 h^2}}\right)^3 \qquad \text{[A2.3]}$$

The contribution of translational motion in G is, per mole, equal to

$$G_t = RT\ln p - RT\left\{\frac{5}{2}\ln T + \ln\left[\left(\frac{2\pi \mathcal{M}}{\mathcal{N}_0 h^2}\right)^{\frac{3}{2}} k^{\frac{5}{2}}\right]\right\} \quad [A2.4]$$

and in entropy, $S_t = \frac{5}{2}R\ln T - R\ln p + R\left\{\frac{5}{2} + \ln\left[\left(\frac{2\pi \mathcal{M}}{\mathcal{N}_0 h^2}\right)^{\frac{3}{2}} k^{\frac{5}{2}}\right]\right\}$.

We can also see that $C_{p,t} = \frac{5}{2}R$ and $C_{v,t} = \frac{3}{2}R$.

Note that the relation giving the entropy runs counter to Nernst's theorem ($T = 0 \Rightarrow S = 0$). This is due to the fact that, at very low temperatures, Boltzmann statistics are no longer applicable and must be replaced by Bose–Einstein or Fermi–Dirac statistics. This area of low temperatures is beyond the remit of this book, which is that of moderate or high temperatures. It is for the same reason that we have assumed equality between the partial specific heat in the mixture and the specific heat of the species in the pure state.

Note once more that the thermodynamic functions are calculated in the tables for a pressure $p = p_r = 10^5\,Pa$.

The contribution due to the internal motion is a function of the structure of the molecule or atom.

A2.3. Internal motions

A2.3.1. *Monatomic species*

The internal contribution in monatomic species, be they ionized or otherwise, is due in the domain of temperature in which we are interested, to the *motion of the electrons* and to the possible jump of those electrons from one orbital to another, so the internal contribution is reduced to the electron contribution: $Q_{el} = \sum_{\ell} g_l \exp(-\varepsilon_l / kT)$. g_l is the static weight of the energy level ε_l, which is identified with the number of quantum states.[1] If J_l is the quantum number relative to the angular moment of the electrons, $g_l = 2J_l + 1$, the energy ε_0 of the first level is supposed to be zero (basic level) and ε_l is generally expressed in terms of frequency $\omega_l\,(\text{cm}^{-1})$ because $\omega_l = hc\,\omega_l$ (h is Planck's constant, and c is the speed of light). Tables give the values of J_l and ω_l. Atomic oxygen, for instance, is characterized by the data in Table A2.1 [ATO 69].

$\omega_l\left(\text{cm}^{-1}\right)$	J_l
0	2
158.5	1
226.5	0
15868	2
13792	0
33768	2
76795	1
86630	7
88630	4

Table A2.1. *Values of the quantum number of the angular moment of the electrons J_l as a function of the frequency ω_l*

Given these data, the partition function becomes

$$Q_{el} = g_0 + g_1 \exp\left(-\varepsilon_1/kT\right) + \ldots$$

The number of terms needing to be taken into account depends on the nature of the plasma and, when temperatures are no higher than 15,000 K, the first terms suffice.

With the first two terms, the entropy, for instance, is equal to

$$S_{el} = R\left\{\frac{\varepsilon_1}{kT}\frac{g_1 \exp\left(-\varepsilon_1/kT\right)}{g_0 + g_1 \exp\left(-\varepsilon_1/kT\right)} + \ln\left[g_0 + g_1 \exp\left(-\varepsilon_1/kT\right)\right]\right\} \quad \text{[A2.5]}$$

1 In practice, the excited electron levels are rarely populated. They represent only a tiny percentage of the total population. It stems from this that in particular cases, only the function Q_{el} is reduced to the sum of a few terms, as the others are zero.

and the enthalpy to

$$H_{el} = RT\frac{\varepsilon_1}{kT}\frac{g_1 \exp(-\varepsilon_1/kT)}{g_0 + g_1 \exp(-\varepsilon_1/kT)}$$

A2.3.2. *Diatomic species*

In addition to the electron contribution seen above, we need to take account of the rotation and vibration of the atoms and of the coupling between rotation and vibration. The partition function will include three terms: $Q_{\text{int}} = Q_r Q_v Q_{el}$. The moment of inertia I with respect to the rotational axis is $I = r_0^2 M_1 M_2/(M_1 + M_2)$, with M_1 and M_2 being the respective masses of the two atoms and r_0 their equilibrium distance. The energy of the internal mode ε_j is given by $\varepsilon_j = j(j+1)h^2/8\pi^2 I = k\,B\,j(j+1)$, with j being the rotational quantum number, which assumes all the integer values for non-symmetrical molecules and the even or odd numbered values for symmetrical molecules.

The *rotational partition function* is of the same form as that written for the different electron levels:

$$Q_r = \sum_{j=0}^{\infty}(2j+1)\exp(-\varepsilon_j/kT) = \sum_{j=0}^{\infty}(2j+1)\exp\left[-\frac{B}{T}j(j+1)\right]$$

Using an Euler–Maclaurin series expansion, we obtain

$$Q_r = \left(\frac{\varphi B}{T}\right)^{-1}\left(1 + \frac{B}{3T} + \frac{B^2}{15T^2}\right) \qquad \text{[A2.6]}$$

when $\varphi = 1$ for non-symmetrical molecules, $\varphi = 2$ for symmetrical molecules.

The value of B is defined by two pieces of spectrographic information about the molecule, B_e and α_e, such that $B = B_e - \alpha_e/2$.

The *contribution due to the vibration* is expressed by a partition function equal to $Q_v = \sum_v \exp(-v\,h\,c\,\omega/k\,T)$, in which v is the vibrational quantum number, setting

$u = hc\omega/kT = 1.43884\,\omega/T = \theta_v/T$, with ω being expressed in cm^{-1} and T in K, we obtain $Q_v = \left(1-e^{-u}\right)^{-1}$, because $\sum_{v=0}^{\infty} x^v = 1 + x + x^2 + \ldots = (1-x)^{-1}$.

Spectrography of the diatomic molecules shows that a certain *coupling between vibration and rotation* exists. The energy taking account of this coupling is written

$$\frac{\varepsilon_{jv}}{hc} = \left(v+\frac{1}{2}\right)\omega_e - \left(v+\frac{1}{2}\right)^2 x_e\omega_e + j(j+1)B_e - j^2(j+1)^2 D_e - \left(v+\frac{1}{2}\right)j(j+1)\alpha_e$$

$\omega_e, x_e, B_e, D_e, \alpha_e$ are spectrographic constants which depend on the molecular structure, e.g. for the molecule HF [KAR 55]: $\omega_e = 4137.25$cm^{-1}, $\omega_e x_e = 88.726$cm^{-1}, $B_e = 20.9456$cm^{-1}, $D_e = 0.002131$cm^{-1}, $\alpha_e = 0.7888$cm^{-1}.

The expression of the partition function includes a double sum of zero at infinity: $Q_{\text{int}} = \sum_{v,j}(2j+1)\exp\left(-\varepsilon_{jv}/kT\right)$. Integrating this double sum leads us to add, to the contributions of the vibration and rotation, that of the coupling equal to

$$\ln Q_{\text{coupl}} = \frac{8\gamma}{u} + \frac{\delta}{e^u - 1} + \frac{2Xu}{\left(e^u-1\right)^2} \qquad [A2.7]$$

with $X = \omega_e x_e / B_e$, $\delta = \alpha_e / B_e$, $\gamma = \left(D_e / 4B_e\right)^{1/2}$, δ being a function of γ and X: $\delta = 6\gamma^{1/2}\left(X^{1/2} - \gamma^{1/2}\right)$.

The calculation of the contribution at the different electron levels is done in the same way as already illustrated with atomic species.

A.2.3.3. *Linear polyatomic species*

With this type of substance, in relation to diatomic species, only the contribution to vibration changes.

The distribution function depends on the number of degrees of freedom – which, in the case a linear molecule, in equal to 3n - 5, with n being the number of atoms. Hence, there are 3n - 5 privileged frequencies, and Q_v is written $Q_v = \prod_{i=1}^{3n-5} \frac{1}{1-e^{-u_i}}$.

The contribution to rotation is identical to that for diatomic molecules, as long the moment of inertia is $I=\sum_{i=1}^{n} M_i x_i^2$, where M_i is the mass of the atom i and x_i is the algebraically-calculated distance from the atom to the center of mass defined by $\sum_{i=1}^{n} M_i x_i = 0$.

A2.3.4. *Nonlinear polyatomic species*

The conclusions are similar to those drawn from the previous section. In this case, the number of degrees of freedom is $3n-6$, so the contribution to vibration is given by $Q_v = \prod_{i=1}^{3n-6} \frac{1}{1-e^{-u_i}}$.

With regard to rotation, the molecule has three moments of inertia I_1, I_2, I_3. In this case, ε_j / kT has three values: $\frac{h^2}{8\pi I_1 kT}, \frac{h^2}{8\pi I_2 kT}, \frac{h^2}{8\pi I_3 kT}$.

If we limit ourselves to the first term of the expansion:

$$Q_r = \frac{1}{\varphi}\left(\frac{8\pi^2 kT}{\hbar^2}\right)^{3/2} \left(\pi I_1 I_2 I_3\right)^{1/2} \qquad \text{[A2.8]}$$

φ being the number of symmetries. This number φ is given for various molecules in Table A2.2.

HOF	$\varphi = 1$
Li_2O	2
LiOH	1
$Li_2O_2H_2$	4
$(LiF)_3$	6
LiOF	1
C_2H_4	4
C_2H_4O	2
CH_3F	3
CH_2F_2	2
NH_2	2

Table A2.2. *Number of symmetries φ for different substances*

These are, very succinctly expressed, the methods used to determine the thermodynamic functions of each species. The heats of formation that play a part in the expression of the internal energy and enthalpy are generally deduced from experiments in thermochemistry. Calculation of these heats of formation on the basis of the bond energies and the molecular structure often yields an imprecise result. However, the technique may be used for an initial approximation for uncommon substances which are difficult to synthesize. In any case, the heats of formation for many substances are not well known as yet, and further experimental efforts need to be made in this domain in order to determine them.

Numerical data relating to the thermodynamic functions and to the heats of formation are to be found in [JAN 69; QUI 50; ROS 09; BER 65]. The number of species dealt with in these tables is sufficient for most applications in thermochemistry.

Appendix 3

Concepts of Turbulence

This appendix deals with turbulent flow, as opposed to laminar (non-chaotic) flow.[1] A laminar flow may become a turbulent flow after a destabilization. Turbulent flows are characterized by their high complexity.[2] Turbulent processes can be modeled using a variety of numerical simulations associated with computational fluid dynamics (CFD).

In section A3.1, we present experimental data on turbulent flows taking place in pipes (an experiment carried out by Reynolds in the 19th Century) or near to solid walls. We can see that the friction coefficients are noticeably altered by the apparition and development of turbulence. Aerodynamics engineers were among the first to calculate turbulent friction coefficients, used to study and design profiles for airplane wings. More specifically, they were applied to evaluate changes in drag and lift, and the occurrence of aerodynamic stalls (the phenomenon of stall can be dealt with, but never completely avoided).

1 As Chassaing [CHA 01] puts it, "The laminar regime presents a high degree of macroscopic 'regularity'. The flow in fact occurs by 'sliding of the fluid layers over one another. The only exchanges between the layers, which result from molecular agitation, lead to diffusion [which is] imperceptible on the scale of the continuum, unlike what we can observe in a turbulent regime."

2 Again according to Chassaing, turbulent agitation manifests itself by unorganized three-dimensional motions which are present, on the scale of the continuous medium, throughout the field of flow.

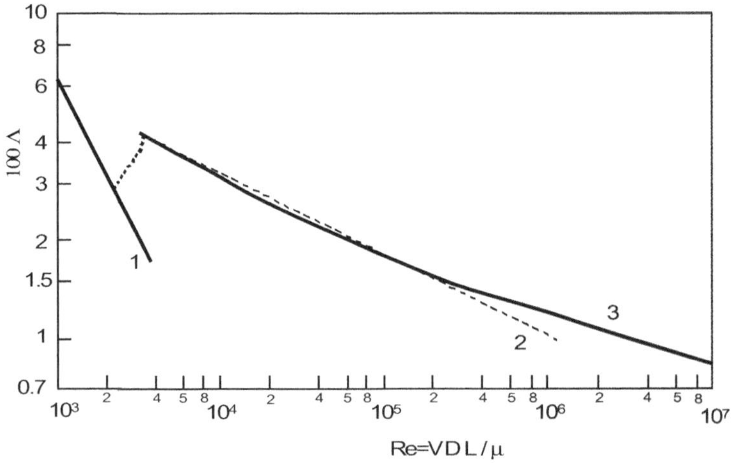

Figure A3.1. *Laminar-to-turbulent transition in a tube (redrawn from the results published by Schlichting [SCH 00]). Curve 1: laminar formula; curve 2: turbulent formula; curve 3: experimental results*

Section A3.2 deals with the apparition of turbulence and its dissipation. We study – analytically, as far as possible – examples of destabilization of laminar flows, which may lead to turbulence if the stresses applied to the system in question are increased. We describe the Rayleigh-Taylor, Kelvin-Helmholtz and Taylor–Couette instabilities, and a flame instability. Once unstable eddies have appeared, they begin to diminish by small-scale dissipation.

The classic turbulence theory (Reynolds-averaged Navier–Stokes or "RANS" theory) is discussed in section A3.3, mainly for flows with a single component. We introduce the average values for all the quantities (compressible and dilatable fluid flows necessitate the use of Favre averages), and turbulent transfer coefficients (similarly to molecular transfer coefficients). Statistical modeling (e.g. using the $k-\varepsilon$ method, which is frequently used in industrial applications) and spectral analysis, which have long been the main tools used for analysis of turbulent flows (often incompressible and non-reactive), are then presented.

The difficulties encountered with the use of statistical methods, and the continually increasing power of computers, have led to attempts to analyze turbulent flows by direct numerical simulation (DNS). Indeed, with most phenomena, turbulent flows are governed by the same fundamental equations as laminar flows, i.e. the Navier–Stokes equations in a pure fluid, but they are characterized by disorganized fluctuations of velocity and variables of state in space and time, with disparities of scale.[3]

However, with DNS, attempting to solve the fundamental equations directly, the costs and the time required soon become prohibitive. An attempt has been made to solve this problem by combining direct numerical simulation (for large eddies) and statistical modeling (for small eddies). This combination of approaches, called LES (*Large Eddy Simulation*), is described in section A3.4.[4]

A3.1. Experimental demonstration

A3.1.1. *Reynolds' experiment*

The existence of a threshold beyond which certain flows exhibit significant and disorganized fluctuations over time has been demonstrated. Reynolds' experiment (1883) relates to the flow of water in a circular tube [REY 83]. A dye is used to help visualize the turbulence. Whilst the Reynolds number $\mathrm{Re} = \rho VD/\mu$ is less than 2000, the regime is laminar, and as we saw earlier, in a laminar regime the head loss coefficient Λ is such that:

$$\Lambda = \frac{D}{\rho V^2/2}\frac{\Delta p}{\Delta \rho} = \frac{64}{\mathrm{Re}} \qquad \text{[A3.1]}$$

After a transition zone, an increase in the Reynolds number leads to a different variation in Λ, even in the absence of roughness of the surface (Figure A3.1):

$$\Lambda \cong 0.3164\,\mathrm{Re}^{-1/4} \qquad \text{[A3.2]}$$

---

3 The distinctive traits on turbulent agitation can be set out in the manner of P. Chassaing: supramolecular size; random behavior; infinite numbers; three-dimensional structure; intermittent presence; rotational kinematics; nonlinear dynamics; dissipative energy [CHA 01]. It then becomes reasonable to introduce statistical methods and to look for only the probabilistic aspects of the solutions, i.e. to view turbulence as a random process [WIL 85].

4 A comparison can be drawn between the computational approaches, distinguishing three levels: RANS, LES and DNS. For DNS, in order to compute all the scales of the Kolmogorov cascade (see sections A3.3.2.3 and A3.3.4), we must have $\mathrm{Re}_t < N^{4/3}$, where Re_t is a turbulent Reynolds number, defined using the integral scale of the turbulence ℓ and the Kolmogorov scale length ℓ_K such that $\mathrm{Re}_t = (\ell_k/\ell)^{4/3}$. This is a less restrictive condition for LES approaches, because $\mathrm{Re}_t < (qN)^{4/3}$ with $q = l_c/\ell_K > 1$, where l_c is the cut-off length scale. Consequently, in terms of cost, LES is significantly more expensive than RANS, but is also more accurate. Yet LES is still less costly than DNS.

A3.1.2. *Viscous flow over a smooth plane plate*

In the case of a plane plate parallel to the average velocity $\vec{\mathbf{V}}$ of the flow, we observe the development of a boundary layer from the leading edge of the plate. Beyond a thickness $\delta(x)$ of this layer, the flow is that of a perfect fluid [BRU 70; DAR 82]. The laminar problem was examined in Chapter 1 of [PRU 12] in section 1.7.2.3.

We shall define the local friction coefficient:

$$C_f(x) = 2\tau(x)/\rho V^2 \qquad \text{[A3.3]}$$

where $\tau(x)$ is the local friction tension, and the average friction coefficient:

$$C_{fm} = \frac{1}{L}\int_0^L C_f(x)\,dx \qquad \text{[A3.4]}$$

Between abscissa values $x = 0$ and x_T, the boundary layer is laminar and we find $C_{fm} = 1.328/\sqrt{\text{Re}}$, with $\text{Re} = \rho VL/\mu$. Beyond x_T, the boundary layer becomes turbulent. In reality, there is no one specific abscissa value x_T, but rather a zone of laminar-to-turbulent transition. If the layer were turbulent from the leading edge, we would find $C_{fm} = 0.074/\text{Re}^{0.2}$.

As the flow is turbulent from x_T, we have the following, where $\text{Re}_T = \rho V x_T/\mu$:

$$C_{fm} = 0.074/\text{Re}^{0.2} - (x_T/L)\left(0.074/\text{Re}^{0.2} - 1.328/\sqrt{\text{Re}_T}\right) \qquad \text{[A3.5]}$$

The results in fact differ because of the extent of the transition zone, and an uncertainty arises about the value of x_T. In addition, with high Reynolds numbers, the formula $\text{Re}^{-0.2}$ is no longer valid, and is replaced by a Schultz-Grunow law in $(\log \text{Re})^{-0.2}$ for the local friction coefficient, so the most appropriate formula is that of Prandtl–Schlichting, which is valid up to values of $\text{Re} \cong 10^9$:

$$C_{fm} = 0.455(\log \text{Re})^{-2{,}58} - 2A\,\text{Re}^{-1}$$

The coefficient A depends on the transition Reynolds number Re_T as shown in Table A3.1 [SCH 00].

Re_T	$\mathbf{10^5}$	$\mathbf{3.\,10^5}$	$\mathbf{5.\,10^5}$	$\mathbf{10^6}$	$\mathbf{3.\,10^6}$
A	150	525	850	1650	4350

Table A3.1. *Dependency of the coefficient A on the Reynolds number*

A3.1.3. *Effect of roughness of the plate*

For a rough plate with homogeneous roughness of thickness *e*, the relative roughness would be $\varepsilon_\delta = e/\delta$, where δ is the thickness of the boundary layer.

With relative roughnesses of less than 1/3, the wall is aerodynamically smooth [BRU 70]. As the thickness δ of the boundary layer is an increasing function of the distance x from the leading edge, it follows that the plate will be rough at the beginning and may then come to appear smooth.

In the rough zone, here we can use Droblenkov's law: $C_f = 0.0139\varepsilon^{1/7}$, where $\varepsilon = e/L$.

A complete, theoretical and numerical study of the instabilities and transitions of sheared flows was performed by Schmid and Henningson [SCH 01].

A3.1.4. *Effect of turbulence on chemical reactivity*

The effects of turbulence can also be seen on the chemical reactions [WIL 75]. The small scales of turbulence are responsible for what we call micromixing in chemical engineering (see Chapter 2).

The following example demonstrates the importance of the micromixing in terms of the progression of a chemical reaction. The reaction in question is the precipitation of barium sulfate by H^+ ions from a stable complex in a basic medium used at the *Laboratoire du génie de la réactivité chimique* (Chemical Reactivity Engineering Lab) in Nancy, France. The species A and B are liquid, as is the intermediary product R; S denotes barium sulfate. We have two competing reactions:

$$A + B \rightarrow R \text{ (very rapid)}; \; R + B \rightarrow S\downarrow \text{ (rapid)} \qquad [A3.6]$$

A drop of B is injected into the liquid A. Two extreme situations may occur, depending on the quality of the turbulent stirring of the liquid A:

1. Effective micromixing on the molecular scale at the injection point of the drop of B into the liquid A. B is always in contact with A, not with R. The precipitate S cannot be formed.

2. Imperfect micromixing. B is surrounded by R; R reacts with B to form the precipitate S [VIL 82; PRU 10].

The amount of S produced enables us to measure the degree of segregation of the mixture, which is greater the more imperfect the micromixing is.

Large eddies do not have a direct influence on the reaction, but are responsible for the residence time of the species in an open reactor (see Chapter 2), for which we would need to determine a residence time distribution t_S.

A3.2. Apparition and damping of turbulence

Numerous situations can lead to instability of a laminar flow. We know that for high Rayleigh numbers [CHA 61], Rayleigh–Bénard rolls become fluctuating [FAU 01; GUY 91]. This phenomenon also occurs at the points of bifurcation between two stable regimes.

The causes of instability are many. In a flow of simple fluid, we examine the Rayleigh–Taylor, Kelvin–Helmholtz and Taylor–Couette instabilities, and we shall touch on hydrodynamic instability of a premixed flame to illustrate a reactive scenario.

The examples presented here demonstrate mechanisms, but the theories rarely yield correct quantitative values [SCH 01]. This is one of the as-yet-unsolved problems in terms of turbulence, and often the only solution to quantitatively determine the boundary between two regimes is still an experimental approach.

A3.2.1. *Instability between two superposed fluids*

For superposed non-viscous fluids, separated by a horizontal surface, with respective density and velocity (ρ_1, U_1), for the lower fluid and (ρ_2, U_2) for the upper fluid, and constant surface tension σ, the amplification factor ω_i is obtained by linearized small-disturbance analysis.

We suppose the fluids to be non-viscous and the layers to be semi-infinite. The velocity potential ϕ obeys the Laplace equation. The disturbance ϕ' can be decomposed into normal modes [LAM 45; GAT 01]. The integration constants are determined using the interfacial conditions. The reference interface being flat and having the equation $z = 0$, the disturbed interface will be described by $z = \eta(x,t)$ in a two-dimensional configuration. The boundary conditions at the interface (Laplace law for the pressure jump caused by the surface tension and conditions of impermeability) are linearized, and the equations for the motion of the fluids can be expressed by the Bernoulli laws, which are also linearized. We obtain a homogeneous system of time-dependent disturbance equations, with the spatial derivatives ϕ' and the function η.

Finally, after elimination of the functions, we obtain the dispersion relation:

$$\left(\omega - k\frac{\rho_1 U_1 + \rho_2 U_2}{\rho_1 + \rho_2}\right)^2 = \frac{\sigma}{\rho_1 + \rho_2}k^3 + \frac{\rho_1 \rho_2}{(\rho_1 + \rho_2)^2}(U_1 - U_2)^2 k^2 + g\frac{\rho_2 - \rho_1}{\rho_1 + \rho_2}k \quad \text{[A3.7]}$$

The amplification factor ω_i, which is the imaginary part of the pulsation ω, can be deduced from the relation:

$$\omega_i^2 = \frac{\sigma}{\rho_1 + \rho_2}k^3 + \frac{\rho_1 \rho_2}{(\rho_1 + \rho_2)^2}(U_1 - U_2)^2 k^2 + g\frac{\rho_2 - \rho_1}{\rho_1 + \rho_2}k \quad \text{[A3.8]}$$

where k is the wavenumber and g is the acceleration due to gravity.

A3.2.1.1. *Rayleigh–Taylor instability*

The Rayleigh–Taylor instability can arise in cases where the surface tension is zero and the velocities of the two fluids are equal ($\sigma = 0, U_1 = U_2$), the upper fluid being the denser ($\rho_2 > \rho_1$). The flat surface is then unstable for any wavenumber (i.e. for any propagation rate).

Conversely, if $\sigma \neq 0$ when always $U_1 = U_2$ and $\rho_2 > \rho_1$, stability analysis gives us a cut-off wavenumber k_c such that $k_c^{\,2} = (\rho_2 - \rho_1)g/\sigma$.

The stability is assured for small wavelengths ($k > k_c$). The maximum value for ω_i occurs with a particular value k_m of the wavenumber (see Figure A3.2), given by

$$k_m{}^2 = (\rho_2 - \rho_1) g / 3\sigma \qquad \text{[A3.9]}$$

A3.2.1.2. *Kelvin–Helmholtz instability*

The Kelvin–Helmholtz instability occurs, with different velocities for the two fluids, and we can assume that g and σ are zero. A cut-off wavenumber

$$k_c = \rho_1 \rho_2 (U_1 - U_2)^2 / \sigma (\rho_1 + \rho_2)^2 \qquad \text{[A3.10]}$$

appears, and for real values of the wavenumber *k*, we have a complex pulsation which is formulated as $\omega = \omega_r + i\omega_i$.

The real part is $\omega_r = k(\rho_1 U_1 + \rho_2 U_2)/(\rho_1 + \rho_2)$ and the imaginary part is given by [A3.8]. We obtain the maximum increase in disturbance for a particular wavenumber (see Figure A3.3) [GAT 01]:

$$k_m = 2\rho_1 \rho_2 (U_1 - U_2)^2 / 3\sigma (\rho_1 + \rho_2)^2 \qquad \text{[A3.11]}$$

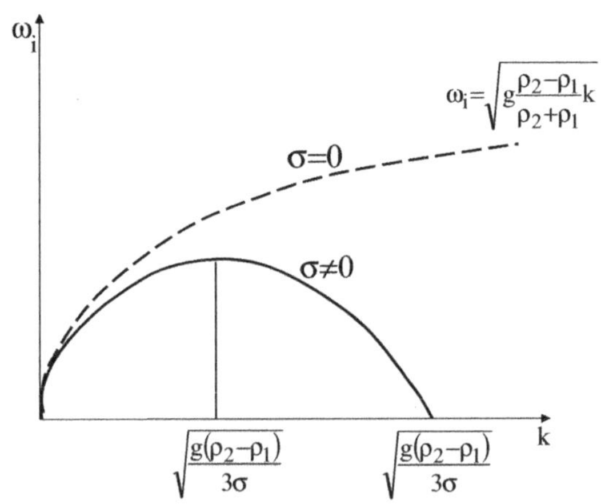

Figure A3.2. *Curves of increasing Rayleigh–Taylor instability in non-viscous fluids when $U_2 = U_1$*

The Kelvin–Helmholtz instability gives rise to rolled structures such as those shown in Figure A3.4a.

It is possible to determine the rate of convection U_c of large structures as indicated below. The stream lines of the flow observed in a referential frame moving at velocity U_c are represented in Figure A3.4b, and exhibit saddle-type structures [PAP 89] similar to those observed for the contra-current diffusion flame and for the stretched plane premixed flame. This observation justifies the local use of these Lagrangian structures in the study and modeling of turbulent flames (Chapter 3).

To determine the velocity U_c of a continuous flow of compressible fluid, we write that the dynamic pressures p_0 of the two flows are equal at the stagnation point, i.e. the ratios $p_0/p=\left[1+(\gamma-1)M^2/2\right]^{\gamma/(\gamma-1)}$ where p is the static pressure, γ the isentropic coefficient and M is the Mach number, are the same. If the Mach numbers are not too high, and if we suppose there to be isentropy, we obtain $\rho_1(U_1-U_c)^2=\rho_2(U_c-U_2)^2$. If U_1 is the highest velocity, we deduce from this that:

$$U_c=U_1\frac{1-U_2/U_1\sqrt{\rho_2/\rho_1}}{1+\sqrt{\rho_2/\rho_1}} \qquad \text{[A3.12]}$$

This formula is no longer valid if the flow does not remain isentropic (shockwaves attached to the structures may appear at high velocities). For $M_c > 0.6$, the mixing layer observed experimentally becomes turbulent and exhibits 3D structures.

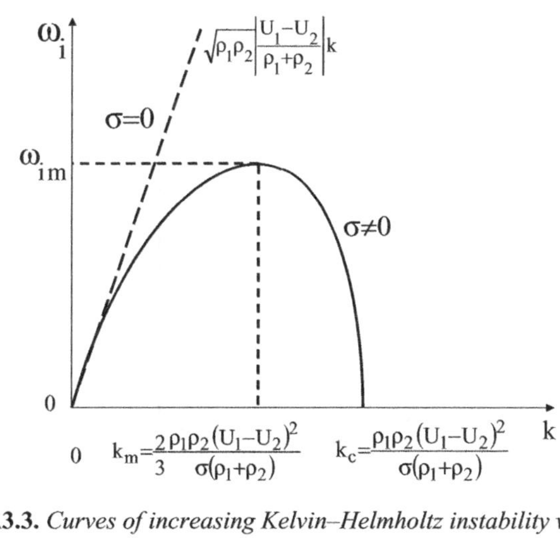

Figure A3.3. *Curves of increasing Kelvin–Helmholtz instability when g = 0*

A3.2.2. *Instability of a fluid between two rotating cylinders*

Now let us consider the flow of a single, incompressible fluid between two rotating cylinders. The case of laminar flow was briefly dealt with in exercise 1.8.6.4 in Chapter 1 of [PRU 12], supposing the outer cylinder to be immobile. In general, the coaxial cylinders (Figure A3.5) have different radii $R_1 < R_2$ and different angular velocities of rotation ω_1 and ω_2 [CHA 61].

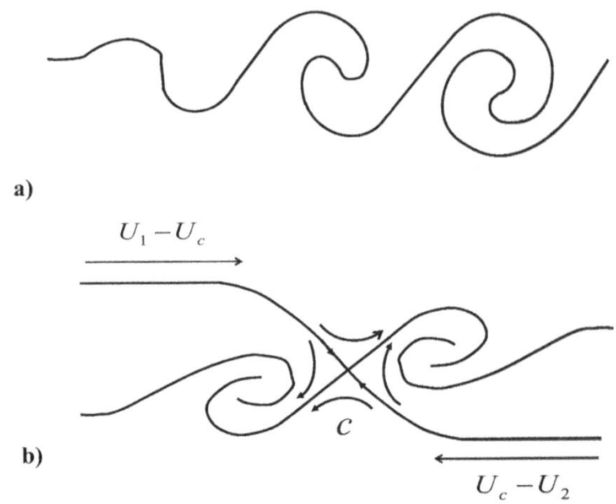

Figure A3.4. *a) Example of a Kelvin–Helmholtz instability which causes rolling at the interface. b) Stream lines in the reference frame with translational velocity* U_c

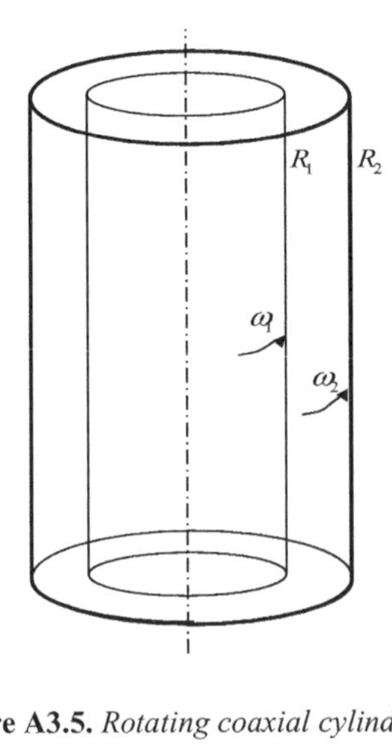

Figure A3.5. *Rotating coaxial cylinders*

A Newtonian viscous fluid is placed between the two cylinders, supposed to be of infinite length.[5] We accept that the axial velocity v_z is zero and look for solutions which give a similar zero value for the radial velocity v_r. We suppose that $\mu = const.$

A3.2.2.1. *Couette flow*

The continuity equation therefore shows that the tangential velocity v_θ is a function of the lone radial coordinate r. The steady solutions found will therefore be axisymmetrical and invariant by axial translational motion.

The momentum equation gives us the relations:

$$-\frac{U_\theta^2}{r^2}+\frac{1}{\rho}\frac{\partial p}{\partial r}=0,\ \frac{1}{\rho r}\frac{\partial p}{\partial \theta}=\nu\left[\frac{1}{r}\frac{\partial}{\partial r}\left(r\frac{\partial U_\theta}{\partial r}\right)-\frac{U_\theta}{r^2}\right],\ \frac{\partial p}{\partial z}=0 \qquad [A3.13]$$

with $\nu=\mu/\rho$ and the boundary conditions $r=R_1 \rightarrow U_\theta=\omega R_1$; $r=R_2 \rightarrow U_\theta=0$. In addition, we shall accept that $\partial p/\partial\theta=0$.

The second equation is written $\frac{\partial}{\partial r}\left[\frac{1}{r}\frac{\partial}{\partial r}\left(rU_\theta\right)\right]=0$. Its solution is of the form: $U_\theta = Ar + B/r$. The integration constants A and B are calculated on the basis of the boundary conditions. We finally obtain:

$$v_\theta=\frac{\omega_2 R_2^2-\omega_1 R_1^2}{R_2^2-R_1^2}r+\frac{R_1^2R_2^2\left(\omega_1-\omega_2\right)}{R_2^2-R_1^2}\frac{1}{r} \qquad [A3.14]$$

The velocity profile obtained does not depend on the viscosity coefficient, which was foreseeable because the equation verified by v_θ is simply $\Delta v_\theta=0$. There is similarity with the steady heat conduction problem, which gives us $\Delta T=0$ and does not include heat conductivity λ.

Hence, there is no viscous laminar boundary layer that can be defined directly on the basis of the velocities profile and μ.

---

5 The results of the study remain valid for cylinders of finite length, at a sufficient distance from the ends of cylinders whose length is sufficiently great in comparison to the radius R_2.

Shear tension: $\tau_{r\theta} = 2\mu \dfrac{R_1^2 R_2^2}{R_2^2 - R_1^2}\dfrac{1}{r^2}$ decreases in r^{-2} and is maximum at the wall of the inner cylinder. The rotation vector is constant and its intensity is $\omega = \dfrac{\omega_2 R_2^2 - \omega_1 R_1^2}{R_2^2 - R_1^2}$. The flow is irrotational if $\omega_2 R_2^2 = \omega_1 R_1^2$.

Finally, the structure of such a flow does not lend itself to mixing between neighboring fluid zones during rotation. We can reasonably estimate that the resulting agitation will not favor mixing in the presence of several chemical species.

A3.2.2.2. *Taylor–Couette instability*

The above solution such that $v_r = v_z = 0$ and $v_\theta = Ar + B/r$ verifying [A3.14] is the only possible solution in the form $\vec{\mathbf{v}} = \vec{\mathbf{v}}(r)$. It is stable as long as the quantity $\dfrac{|B|}{\nu} = \dfrac{R_1^2 R_2^2}{\nu}\left|\dfrac{\omega_1 - \omega_2}{R_2^2 - R_1^2}\right|$ is small.[6]

For a sufficiently large value of $|B|/\nu$, we witness the apparition of the Taylor instability [TAY 23; CHO 92; GOD 98], which gives rise to a new solution in the form of parallel rollers. This solution is always steady and axisymmetrical, but it is no longer invariant with axial translational motion.

These Taylor vortices corresponding to a solution $\vec{\mathbf{v}} = \vec{\mathbf{v}}(r,\theta,z)$ or $\vec{\mathbf{v}} = \vec{\mathbf{v}}(r,z)$ becomes unstable in turn when we increase $|B|/\nu$ by increasing the relative speed of rotation of the cylinders. We then see the emergence of corrugated vortices corresponding to a periodic time-dependent regime [JOS 76].

The study of these instabilities is complex. We can envisage steady solutions with axial symmetry to the linearized equations in terms of *(u, v* and *w)*, with these

6 The Rayleigh criterion, based on equilibrium between centrifugal force due to rotation and centripetal force due to the pressure gradient, states that the flow will be stable if $d\left|\omega r^2\right|/dr > 0\ \forall r$. In particular, stability would imply that $\left|\omega_1 R_1^2\right| < \left|\omega_2 R_2^2\right|$, which is not verified for $\omega_2 = 0, \omega_1 \neq 0$, as in exercise 1.8.6.4 [PRU 12] where the flow would be unstable at all points. However, we need to take account of the viscous forces which tend to stabilize the flow and cause the apparition of thresholds, as shown by the linearized analysis below and the experimental results.

quantities being defined as disturbances in velocity in relation to the Couette flow: $p = p(r) + p(r,z), v_r = u(r,z), v_\theta = v_\theta(r) + v(r,z),\ v_z = w(r,z)$.

The equations in the steady regime, after linearization and setting of $\mathcal{L}(f) = \frac{\partial}{\partial r}\left(\frac{1}{r}\frac{\partial(rf)}{\partial r}\right) + \frac{\partial^2 f}{\partial z^2}$, become

$$\begin{cases} \frac{1}{r}\frac{\partial(ru)}{\partial r} + \frac{\partial w}{\partial z} = 0, \quad \frac{1}{\rho}\frac{\partial p}{\partial r} = \nu\,\mathcal{L}(u) + 2\left(A + \frac{B}{r^2}\right)v, \\ 2Au = \nu\,\mathcal{L}(v), \quad \frac{1}{\rho}\frac{\partial p}{\partial z} = \nu\left[\frac{1}{r}\frac{\partial}{\partial r}\left(r\frac{\partial w}{\partial r}\right) + \frac{\partial^2 w}{\partial z^2}\right] \end{cases} \qquad [A3.15]$$

By eliminating w and p in the system [A3.15], we find

$$2\left(A + B/r^2\right)\partial^2 v/\partial z^2 + \nu\,\mathcal{L}^2(u) = 0, 2Au = \nu\,\mathcal{L}(v) \qquad [A3.16]$$

with the boundary conditions $r = R_1, R_2 \rightarrow u = \partial u/\partial r = v = 0$.

We then seek solutions of the type $u = \hat{u}(r)\cos(kz), v = \hat{v}(r)\cos(kz)$.

The above system becomes:

$$2\left(A + B/r^2\right)k^2\hat{v} + \nu\,L^2(\hat{u}) = 0, 2A\hat{u} = \nu\,L(\hat{v}) \qquad [A3.17]$$

with $\mathrm{L}(\hat{f}) = \frac{d}{dr}\left(\frac{1}{r}\frac{d(r\hat{f})}{dr}\right) - k^2$ and $r = R_1, R_2 \rightarrow \hat{u} = d\hat{u}/dr = \hat{v} = 0$.

When the radii R_1 and R_2 are close, we can follow Chandrasekhar's approach [CHA 61] and set $d = R_2 - R_1, \mu = \frac{\omega_2}{\omega_1}, \zeta = \frac{r - R_1}{d}, a = k\,d, \alpha = \mu - 1,\ D = \frac{d}{d\zeta}$ and

$$T = -\frac{4A\omega_1 d^4}{\nu^2}, \hat{u} = \frac{2d^2\omega_1 a^2\hat{U}}{\nu}$$

We then obtain

$$\left(D^2-a^2\right)^2\hat{U}=\left(1+\alpha\zeta\right)\hat{v},\left(D^2-a^2\right)\hat{v}=-Ta^2\hat{U} \qquad [A3.18]$$

where $\zeta=0,1\rightarrow\hat{U}=D\hat{U}=\hat{v}=0$.

The solution to system [A3.18] can be found by expressing $\hat{v}$ in the form $\hat{v}=\sum_{m=1}^{\infty}C_m \sin m\pi\zeta$. Following a fairly complex calculation involving an approximation, we find that the first Taylor rollers appear as soon as T surpasses the critical value $T_c=\dfrac{3430}{1+\mu}$ obtained for a = 3.12.

In particular, the difficulty in solution lies in the fact that the first equation in the system includes the term $1+\alpha\zeta$.

Serrin finds a slightly different solution, replacing this term with $\dfrac{\mu+1}{2}$ (we then set $r^2=R_1 R_2$). The differential system is then linear with constant coefficients.

The solutions to the problem are also known *for having very different radii R_1 and R_2.*

Figure A3.6 gives a representation in the plane of the limits of stability deduced from linearized theory and experimental points.

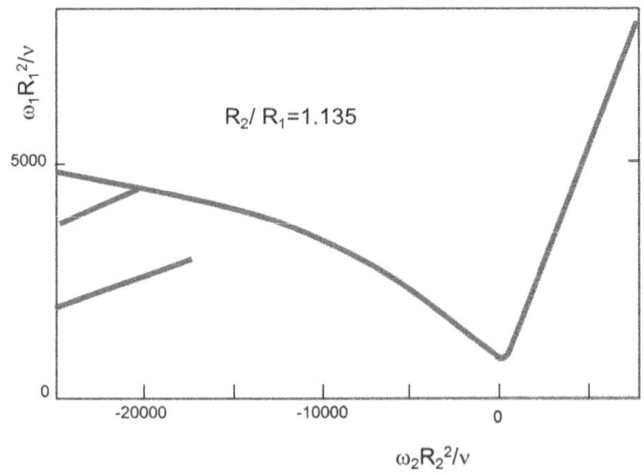

Figure A3.6. *Overview of the theoretical and experiment results found by Joseph (redrawn from [JOS 76])*

A3.2.2.3. *Laminar to turbulent transition*

Beyond the critical value, we may see the appearance of other instabilities such as corrugated vortices. These no longer exhibit axial symmetry, and the fields of velocities and pressures now depend on r, θ, z, t. Spiral vortical flows, when the cylinders are turning in opposite directions, and progressive waves can also be observed. The type of flow obtained depends on the initial disturbance [GOS 96]. In addition, turbulence occurs with high values of the relative angular velocity of the two cylinders [GOL 75].

Figure A3.7 [AND 86; TRI 88] illustrates the complexity of the situations encountered experimentally.[7] Bifurcations (see Appendix 4 and [BER 84], for instance) are numerous. The Reynolds numbers on both axes are not the same as in Figure A3.6. Indeed, we have the definitions

$$\mathrm{Re}_i = R_1 (R_2 - R_1)\omega_1 / \nu,\ \mathrm{Re}_o = R_2 (R_2 - R_1)\omega_2 / \nu$$

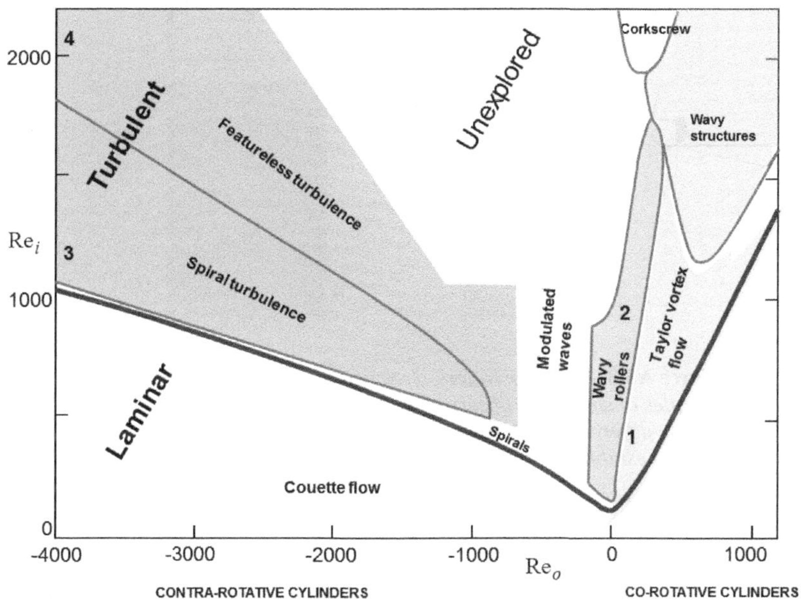

Figure A3.7. *Simplified diagram of the regimes of flow between two concentric cylinders with independent rotations (based on the experiments of Andereck et al. [AND 86]). The lines indicate the transition boundaries which are difficult to establish by mere visual observation. The numbers refer to the examples in Figure A3.8*

7 Further details are given in Figures 2 and 18 in the article by Andereck *et al.* [AND 86] for contra-rotative or co-rotative cylinders.

Figure A3.8 shows:

1. Taylor rollers.
2. Schematized corrugated rollers.
3. A case of spiral instability.
4. Another case of established turbulence.

A number of authors have successfully solved a certain number of these complex situations by theory and numerical calculation.

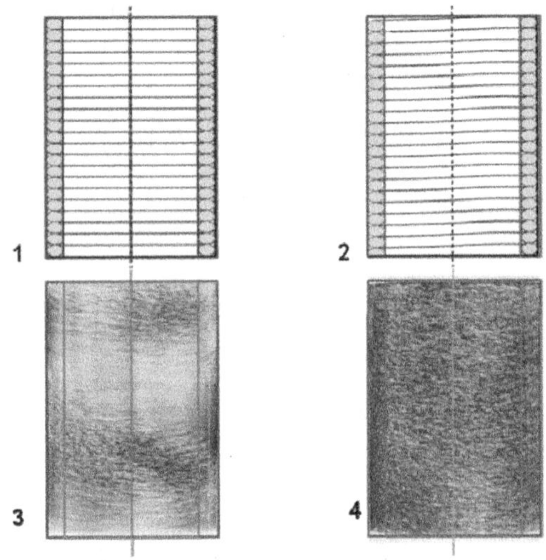

Figure A3.8. *1. Taylor rollers; 2. Schematized corrugated rollers. Examples of structures observed experimentally at high velocities of differential rotation; 3. A case of spiral instability. 4. A case of established turbulence (based on [AND 86], see Figure A3.7)*

A3.2.2.4. *Conclusion about the flow between two rotating cylinders*

In the case of a mixture of several species, and unlike in the case of laminar Couette flow invariant by axial translational motion shown in section A3.2.2.1, beyond the stability threshold the differentiated rotating motion influences the chemical processes, favoring the mixing of the species. This is manifest for turbulent flows, as we shall later confirm.

We shall limit our presentation of the flows between two rotating coaxial cylinders to these results. Various authors have also discussed the case of an axial flow between such cylinders. In the presence of a radial flow resulting from blowing or suction at the walls, we can also predict the existence of instabilities that will give rise to vortices.

A3.2.3. *Instability of a premixed flame*

The premixed flames under consideration here result from the ignition of a gaseous mixture of oxidizer and fuel reacting with a high activation energy. On a large scale, the flame is comparable to a surface across which sudden changes in density, temperature, concentrations and velocity take place. There are various regimes of premixed combustion as explained in Chapter 3. Here, we shall only examine deflagration waves, i.e. slow flames at quasi-uniform pressure (section 3.1.2, as opposed to detonation waves, which are characterized by a pressure jump, similar to what we saw with shockwaves, but all the greater because of the release of reactive energy).

Instability of a premixed flame can easily be obtained by varying the richness of a Bunsen burner flame. Beyond a critical value of the chemical composition of the mixture, the cone is instantly transformed into a polyhedron, as seen in Figure A3.9 [DUR 87]. The spontaneous appearance of cells is observed only with mixtures that are rich in heavy hydrocarbons (propane) or which have only a small fraction of light fuels (hydrogen), i.e. when the species that limits the reaction is the lightest of those in the mixture. Indeed, the flame can exhibit various types of instability (for example, see [PEL 82]).

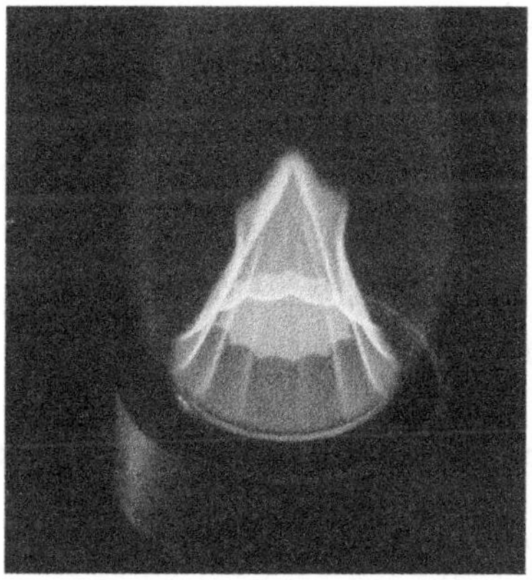

Figure A3.9. *Instability of a premixed propane/air flame. The laser-plane cross-section of the profile of the stable flame is circular. The instability lends it a polygonal form (photo: Laboratoire d'Aérothermique, CNRS, Personal communication)*

Below, we present two types of instability in a laminar premixed flame: hydrodynamic instability and thermal diffusive instability.

A3.2.3.1. *Hydrodynamic instability of a laminar plane premixed flame*

Hydrodynamic instability of a laminar plane premixed flame is known as Darrieus–Landau instability [BOR 00; LAN 71]. Here, we shall study the stability of a plane flame front under the influence of infinitesimal disturbances. The flame of interest is reduced to a discontinuity, as described in section 3.1.1, and is a deflagration wave.

We take the undisturbed plane of discontinuity to be the plane *yz*, the velocities of the gases being oriented in a positive direction along the *x* axis: v_1 for unburned gas and v_2 for the burned gases. We have $\rho_1 v_1 = \rho_2 v_2$ and $\rho_2 < \rho_1$ (because of the heat release), as well as $p_1 = p_2$ (the squared velocity terms are negligible).

The disturbance equations on each side (1 and 2) of the discontinuity are those of non-viscous incompressible fluids:

$$\begin{cases} \vec{\nabla} \cdot \vec{\mathbf{v}}' = 0 \\ \partial \vec{\mathbf{v}}'/\partial t + \left(\vec{\mathbf{v}} \otimes \vec{\nabla}\right) \cdot \vec{\mathbf{v}}' + \left(\vec{\nabla} p'\right)/\rho = \vec{\mathbf{0}} \end{cases} \qquad [A3.19]$$

The disturbed flame surface is defined by $x = \eta\ (y,\ t)$ (two-dimensional case shown in Figure A3.10). On the discontinuity surface (i.e. at $x = 0$), the following conditions must be satisfied:

– pressure continuity equation $p'_1 = p'_2$;

– the condition of continuity of the tangential component of the velocity at the interface: $v'_{1y} + v_1\, \partial\eta/\partial y = v'_{2y} + v_2\, \partial\eta/\partial y$;

– the condition for the normal velocity of the gas at the discontinuity interface to be unchanged: $v'_{1x} - \partial\eta/\partial t = v'_{2x} - \partial\eta/\partial t$.

In the region $x < 0$ (unburned gas, region 1), the solution to [A3.19] is

$$\begin{cases} v'_{1x} = A e^{(iky - kx - i\omega t)},\ v'_{1y} = iA\, e^{(iky - kx - i\omega t)}, \\ p'_1 = A \rho_1 \left(i\omega/k - v_1\right) e^{(iky - kx - i\omega t)} \end{cases} \qquad [A3.20]$$

In the region $x > O$ (combustion products, region 2), we can write [LAN 71]:

$$\begin{cases} v'_{2x} = B e^{(iky - kx - i\omega t)} + C e^{(iky + i\omega x/v_2 - i\omega t)} \\ v'_{2y} = -iB e^{(iky - kx - i\omega t)} - (\omega k/v_2) C e^{(iky + i\omega x/v_2 - i\omega t)} \\ p'_2 = -B \rho_2 (i\omega/k + v_2) e^{(iky - kx - i\omega t)} \end{cases} \quad [A3.21]$$

If we set $\eta(y,t) = D e^{i(ky - \omega t)}$, $\omega = \omega_r + i\omega_i$, a solution will be unstable if $\omega_i > 0$. We obtain four homogeneous equations for the coefficients A, B, C and D. A simple calculation gives the condition for these equations to be compatible – i.e. the following dispersion equation:

$$(v_1 + v_2)\omega_i^2 + 2kv_1v_2\,\omega_i + v_1v_2k^2(v_2 - v_1) = 0 \quad [A3.22]$$

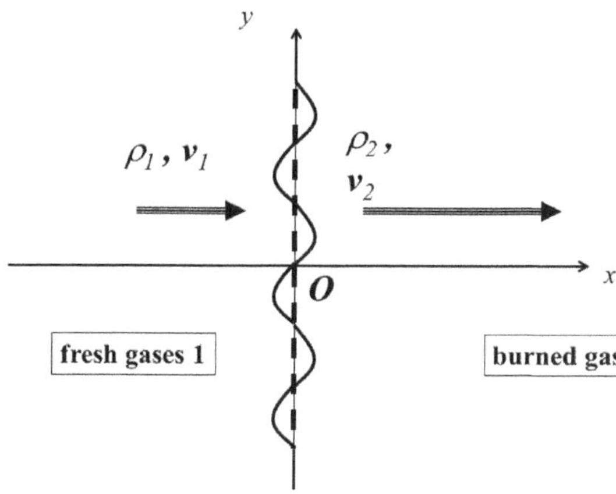

Figure A3.10. *Plane flame front with infinitesimal disturbances*

Here, $v_1 < v_2$ because $\rho_2 < \rho_1$ (as we stated earlier: because of the significant temperature increase during the course of combustion) and $\rho_1 v_1 = \rho_2 v_2$, the roots are real and of opposite signs and the reference motion is therefore unstable.

A3.2.3.2. *Thermal diffusive instability of a laminar premixed flame*

Now let us consider a reactive mixture of dilute species *A* and *B* in a neutral gas, where the composition is far from stoichiometric, and the lighter of the two species is also the rarer [CLA 78]. In this case, the progression of the reaction only has a noticeable effect on species A. The diffusion coefficient is then large (see Chapter 2 of [PRU 12]) because of the low molar mass, and the Lewis number $Le_A = \lambda / \rho c_p D_A$ is therefore small in relation to the Lewis number of species *B* (the heavier species).

The heat diffusivity $\kappa = \lambda / \rho c_p$ is independent of the proportions of *A* and *B* because of the significant dilution of the mixture.

The rate of combustion of a plane laminar premixed flame is clearly defined (see Chapter 3). Locally, a stable Bunsen burner flame is comparable to such a plane flame. Suppose that a local disturbance causes a strain which manifests itself by pits and studs (Figure A3.11). The transversal diffusion of species *A* tends to decrease the concentration C_A at a point *M* on the reaction surface situated in a pit. On the other hand, the heat from the burned gases diffuses inwards. Nevertheless, if Le_A is small, the effect of the diffusion of species *A* predominates, and the reaction rate decreases still further if the temperature at *M* increases, which leads to a decrease in the velocity of the combustion wave, i.e. the celerity of the flame in relation to the fresh gases. The zone in question will become increasingly deeply inset. The opposite happens at point *N* along the section *y'y*. The flame exhibits instability.

If the Lewis number were greater than 1, the effects of heat conduction would be greater than those of molecular diffusion, which would lead to stability.

These instability effects give rise to thermal diffusive autoturbulence by amplification of the defects. If the flow in front of the fresh gases is already turbulent, the two phenomena will interact.

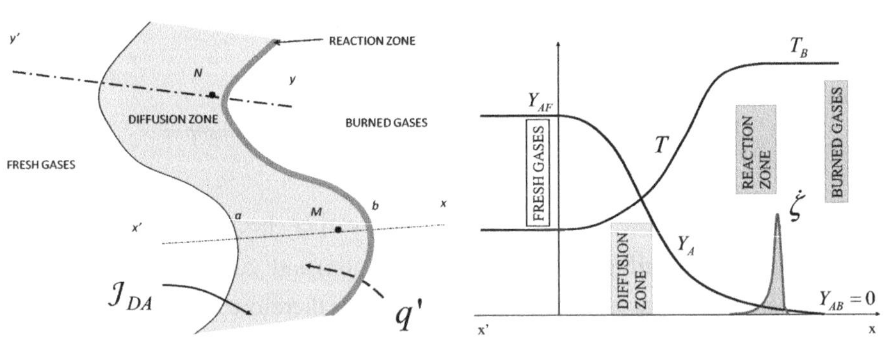

Figure A3.11. *Thermodiffusive instability of a laminar flame: flame geometry on the scale of the thickness of diffusion; profiles of temperature, concentration and reaction rate*

The motion of pleated turbulent flames was the subject of research some four decades ago [WIL 70]. Clavin and Joulin [CLA 83] showed that in the context of certain hypotheses, a single scalar quantifying the stretching of the flame front locally controls the shape and dynamics of the flame (section 3.2.4, [PRU 12]). This scalar comprises two terms which represent the contribution of the geometry of the flame (the curvature of the flame front, advancing at a determinate normal speed) and that of the inhomogeneity of the flow, characterized by the strain rate tensor.

Law, Jomaas and Bechtold [JOM 07; LAW 05] recently studied the appearance of cellular instabilities in the expansion of spherical flames (see Figure 3.1 in Chapter 3 of [PRU 12]).

A3.2.4. *Damping of turbulence*

We have presented a number of mechanisms of instability which could give rise to turbulence. In what follows, we shall see that the turbulent structures evolve by energy exchange. These exchanges may be fairly inertial and dissipative in nature.

The small structures are generally damped by molecular interaction (viscous diffusion, diffusion of matter or conduction). In a pure fluid, we can consider the slowing of the angular rotation of the small vortices under the influence of viscosity. An example of the damping of a plane vortex was discussed in Chapter 1 (exercise 1.8.6.5, [PRU 12]), for which we find

$$\vec{\omega} = \frac{\Gamma}{8\pi\nu t} e^{-\frac{r^2}{4\nu t}} \vec{k} \qquad \text{[A3.23]}$$

where Γ is the initial circulation, ν the kinematic viscosity and $\vec{\mathbf{k}}$ the unitary vector normal to the plane of motion. Note, however, that the concept of a turbulent structure can certainly not be reduced to that of a free vortex. Roughly speaking the turbulent vortices are coherent and interdependent structures which are frequently used in turbulence theory in relation with spectral analysis.

Also, the turbulent structures under discussion here are generally three-dimensional, and a more appropriate model to describe them would be the Burgers vortex [BUR 48], defined by the equations

$$\begin{cases} v_r = -\sigma r \\ v_\theta = \dfrac{\Gamma}{2\pi r}\left(1 - e^{-\sigma r^2/2\nu}\right) \\ v_z = 2\sigma z \end{cases} \qquad \text{[A3.24]}$$

where σ denotes the strain rate and Γ the circulation of the vortex. The value $\delta = \sqrt{\nu/\sigma}$ characterizes the size of the core of this vortex. Better still, we use the concept of the Burgers vortex layer to represent the damping of the turbulence.

The right coherent structure to choose to model the turbulence depends on the type of flow studied, and it is not always easy to identify [JEO 95].

Diffusion of species also plays a part in reactive flows, as it does in chemical reactivity. To illustrate this, we can – as many authors do [MAR 77; MAR 85; KAR 86] – consider a plane vortex damped by viscosity in the presence of a combustion interface between oxidizer and fuel, i.e. a diffusion flame. We find that the characteristic radius of the viscous vortex is $r^* \approx \sqrt{\nu t}$, whereas the radius of the core of burned gases evolves differently. For instance, $r^* \approx \sqrt{\Gamma^{2/3} D^{1/3} t}$, asymptotically if $\mathrm{Re}\sqrt{Sc} > 50$ [MAR 77].

The dynamics of the interactions between the premixed flame and the vortices has also been studied [LAV 89; REN 00].

A3.3. Classic turbulence (RANS model)

Every characteristic value of the fluid, satisfying the balance equations[8] established in [PRU 12], is fluctuating (see Figure A3.12). When using the RANS (*Reynolds Averaged Navier–Stokes Simulation*) method, we consider local averages for the values or their moments, and then establish averaged balance equations for these quantities. These balances involve source terms and flux densities. Simulations are necessary to ensure the closure of the system and to solve it numerically. The presence of multiple scales at each point and at all times necessitates an appropriate analysis in the physical space and/or in the Fourier transform space (spectral analysis). Below, we present the quantities in question, the equations and the conventional closure techniques.

A3.3.1. *Turbulent transfer and chemical kinetics coefficients*

Here we give as simple a presentation as possible of statistical processing of fluid turbulence.

8 In the case of a single component, these are the *Navier–Stokes* equations.

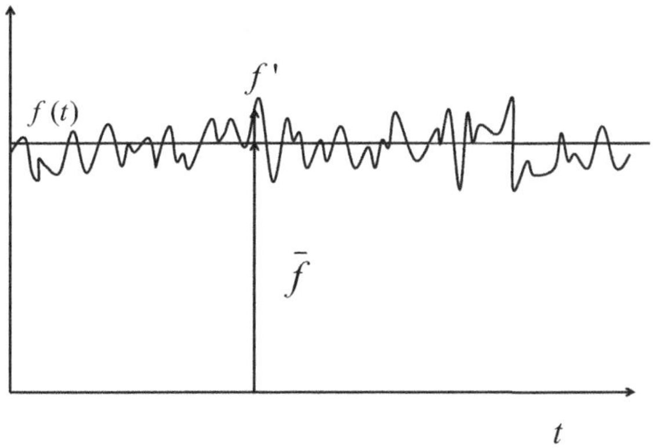

Figure A3.12. *Turbulent value f and its average and fluctuating parts*

A3.3.1.1. *Average quantities*

Reynolds averages

Consider a parameter $f(\vec{\mathbf{x}},t)$ of the flow (pressure, temperature, velocity, concentration, etc.). We can define the average value:

$$\overline{f}(\vec{\mathbf{x}},t)=\frac{1}{T}\int_{\tau=t-T/2}^{\tau=t+T/2} f(\vec{\mathbf{x}},t)\,d\tau \qquad \text{[A3.25]}$$

where T is the sampling delay, chosen to be sufficiently large so that $\overline{f}$ is independent of it. The parameter f then exhibits an average value $\overline{f}$ and a random part f' (Figure A3.12).

We have $f=\overline{f}+f'$, and we suppose the Reynolds axioms to be satisfied:

$$\begin{cases}\overline{f'}=0, \quad \overline{\overline{f}}=\overline{f}, \quad \overline{kf}=k\overline{f}, \quad \text{if } k=const.\\ \overline{fg}=\overline{f}\,\overline{g}+\overline{f'g'}, \quad \overline{\partial f/\partial s}=\partial\overline{f}/\partial s\\ \overline{\int f\,ds}=\int\overline{f}\,ds\end{cases} \qquad \text{[A3.26]}$$

A coefficient of correlation can be defined between f and g

$$R_{fg}=\frac{\overline{f'g'}}{\sqrt{\overline{f'^2}\,\overline{g'^2}}} \qquad \text{[A3.27]}$$

For two points M_1 and M_2 at a distance of ξ, with respective velocities u_1 and u_2 at time *t*, we define a correlation coefficient:

$$R(\xi)=\overline{u'_1 u'_2}\Big/\sqrt{\overline{u'^2_1}}\sqrt{\overline{u'^2_2}} \qquad \text{[A3.28]}$$

and a correlation length L giving the average dimension of the vortices:

$$L=\int_0^{\infty} R(\xi)\,d\xi \qquad \text{[A3.29]}$$

If $R(\xi)$ is independent of the direction of $M_1M_2=\xi=|\vec{\xi}|$, the turbulence is said to be *isotropic*. In addition, the turbulence is *homogeneous* if this property does not depend on the position.

Favre averages

If the fluid in question exhibits significant variations in density (simple compressible fluid or combusting mixture), we generally use the Favre averages. These weighted averages help to greatly simplify the writing of the turbulent balance equations [FAV 58; FAV 71].

The Favre average $\tilde{f}$ of a value $f(\vec{x},t)$ is defined on the basis of the conventional average of the product ρf. We simply write

$$\tilde{f}=\overline{\rho f}/\bar{\rho} \qquad \text{[A3.30]}$$

By decomposing f into $f=\tilde{f}+f"$, we find $\tilde{f"}=0$ and $\overline{\rho f"}=0$.

The main advantage of these Favre averages is that they simplify the formulation of the products involving density.

Indeed, we can compare the development of the conventional average of the product $\rho u f$ which gives four terms, to that of the Favre average which has only two, and has the same structure as the conventional average: $\overline{uf}=\bar{u}\,\bar{f}+\overline{u'f'}$.

$$\begin{cases}\overline{\rho u f}=\bar{\rho}\bar{u}\,\bar{f}+\bar{\rho}\overline{u'f'}+\overline{\rho u'}\,\bar{f}+\overline{\rho f'}\,\bar{u}+\overline{\rho' u' f'}\\ \overline{\rho u f}=\bar{\rho}\tilde{u}\,\tilde{f}+\bar{\rho}\widetilde{u"f"}\end{cases} \qquad \text{[A3.31]}$$

Certain difficulties that arise in finding the Favre average of the terms of diffusive transport are irrelevant for turbulence with large Reynolds numbers, because in this case, these terms are negligible [PET 00].

A3.3.1.2. *Turbulent exchange coefficients*

Turbulence causes the apparition of flux terms which are added to the molecular fluxes of the different values involved because of the transfer phenomena. The balance equation:

$$\frac{\partial(\rho f)}{\partial t}+\vec{\nabla}\cdot\left(\rho f\,\vec{\mathbf{v}}+\vec{\mathcal{J}}_F\right)=\dot{W}_F \quad [A3.32]$$

by passage through the mean, *in the incompressible case*, becomes

$$\rho\partial\bar{f}/\partial t+\rho\vec{\nabla}\cdot\left(\bar{f}\,\bar{\vec{\mathbf{v}}}\right)+\vec{\nabla}\cdot\overline{\vec{\mathcal{J}}_F}+\rho\vec{\nabla}\cdot\overline{f'\vec{\mathbf{v}}'}=\overline{\dot{W}_F} \quad [A3.33]$$

Thus, the momentum equation brings the Reynolds tensor into play:

$$\rho\overline{\vec{\mathbf{v}}'\otimes\vec{\mathbf{v}}'} \quad [A3.34]$$

which is added to the viscous pressures to give the sum of the molecular and turbulent viscosities:

$$-2\mu\vec{\vec{\mathbf{D}}}+\rho\overline{\vec{\mathbf{v}}'\otimes\vec{\mathbf{v}}'} \quad [A3.35]$$

A simplifying hypothesis is to accept that the turbulent fluxes are proportional to the corresponding generalized forces. Thus, here, we would have

$$\rho\overline{u'v'}=-K_u\,\partial\bar{u}/\partial y \quad [A3.36]$$

where K_u is a turbulent exchange coefficient. Unlike molecular transfer coefficients, turbulent exchange coefficients depend not only on the local state of the fluid but also on the characteristics of the flow. It is rare that they can be deemed to be constant. In the case of a turbulent Poiseuille flow, we would accept, for instance, that K_u is a function of y. Thus, we are dealing with semi-empirical coefficients, whose field of validity is generally very restricted, but which exhibit the advantage of ensuring the closure of the system of equations and giving qualitative indications.

Thus, we define the three *turbulent exchange coefficients* $\mu_t, \lambda_t/\overline{c_p}$ and ρD_t such that

$$\begin{cases} \overline{(\rho\vec{\mathbf{v}})'\otimes\vec{\mathbf{v}}'} = -\mu_t \vec{\nabla}\otimes\vec{\mathbf{v}} \\ \overline{(\rho\vec{\mathbf{v}})'h'} = -\lambda/\overline{c_p}\,\vec{\nabla}\overline{h} \\ \overline{(\rho\vec{\mathbf{v}})'Y_j'} = -\rho D_t \vec{\nabla}\overline{Y_j} \end{cases} \quad [A3.37]$$

These relations are also valid in compressible fluid. We can therefore introduce turbulent Schmidt- and Prandtl numbers:

$$Sc_t = \mu_t/\rho D_t\,,\ \mathrm{Pr}_t = \mu_t\overline{c_p}/\lambda \quad [A3.38]$$

By analogy with the kinetic theory of gases, we can speak of a mixture length (similar to the mean free path). For instance, consider a transfer of value F in direction y. We agree that the transfer takes place by random jets of length ℓ_F' in direction Oy, and we suppose that these jets carry with them the average value $\overline{f}(y-\ell_F')$ from their level of origin. That value is preserved during transit, after which it suddenly regains equilibrium with the average value of the destination level $\overline{f}(y)$. At the abscissa value y, the fluctuations f' will therefore be:

$$f' = \overline{f}(y-\ell_F') - \overline{f}(y) \cong -\ell_F'\partial\overline{f}/\partial y \quad [A3.39]$$

Hence, if we take account of all possible random jets and find the average:

$$\overline{(\rho\vec{\mathbf{v}}')f'} = -\overline{(\rho\vec{\mathbf{v}}')\ell_F'}\,\partial\overline{f}/\partial y \quad [A3.40]$$

then we have

$$K_F = -\overline{(\rho\vec{\mathbf{v}}')\ell_F'} \quad [A3.41]$$

Suppose ρ to be constant and the fluctuations u' to be of similar magnitude to the fluctuations v'. If we set $f = u$, we have

$$\begin{cases} u' \cong -\ell_u'\partial\overline{u}/\partial y \cong v' \\ K_u = \mu_t = \overline{(\rho\vec{\mathbf{v}}')\ell_u'} = \rho(\ell_u')^2\,\partial\overline{u}/\partial y \end{cases} \quad [A3.42]$$

If we set $L_P^2 = \overline{\ell_u'^2}$, this defines Prandtl's mixing length L_P, which can be interpreted as the size of the most effective vortices to perform mixing.

A3.3.1.3. *Turbulence and chemical kinetics*

Let us now turn out attention to the turbulent terms due to the chemical kinetics. Consider the chemical reaction $A+B\rightarrow C$.

The rates of species production are

$$\dot{W}_j = \nu_j \mathcal{M}_j k\rho Y_A Y_B / \mathcal{M}_A \mathcal{M}_B = \pm k_j Y_A Y_B \qquad \text{[A3.43]}$$

with the sign (+) for $j = C$ and the sign (–) for $j = A$ or B. For the average, we obtain

$$\overline{\dot{W}_A} = -k_1\left(\overline{Y_A}\,\overline{Y_B} + \overline{Y_A{}'Y_B{}'}\right) - \overline{Y_A}\,\overline{k_1{}'Y_B{}'} - \overline{Y_B}\,\overline{k_1{}'Y_A{}'} - \overline{k_1{}'Y_A{}'Y_B{}'} \qquad \text{[A3.44]}$$

We note that, even when $k_1{}'$ is negligible[9], the average production rate is the sum of two terms, and may therefore be very different from the production rate calculated using the averaged concentrations:

$$\overline{\dot{W}_A} = -k_1\left(\overline{Y_A}\,\overline{Y_B} + \overline{Y_A{}'Y_B{}'}\right) \qquad \text{[A3.45]}$$

By introducing a characteristic diffusion time τ_D and a chemical time τ_c we can show that:

a) if $\tau_D << \tau_c$, i.e. if the reactions are slow or the system is perfectly stirred, the turbulent chemical term is negligible in comparison to the turbulent diffusion term. Then, $\overline{Y_A{}'Y_B{}'} \cong 0$ and

$$\overline{\dot{W}_A} = -k_1\overline{Y_A}\,\overline{Y_B} \qquad \text{[A3.46]}$$

b) if $\tau_c << \tau_D$, the chemical reaction is very fast in comparison to the phenomenon of diffusion. The species are poorly mixed, and in extreme cases, never come into contact with one another. A point in space will alternately experience the presence of species *A* or *B*, but never both at once. The result of this is that the production rate is null, which implies that

$$\overline{Y_A}\,\overline{Y_B} = -\overline{Y_A{}'Y_B{}'} \qquad \text{[A3.47]}$$

9 This is possible if the fluctuation in temperature is negligible, which is certainly not the case in combustion, where the variations in temperature due to the heat release are generally far from negligible. If we neglect $k_1{}'$, it is because we are interested primarily in the effects of mixing.

c) between these two extreme cases, the term of correlation $\overline{Y_A'Y_B'}$ has an absolute value between 0 and 1.

Note that this is a drastically simplified presentation of turbulent reactive mixtures, hinging on the mixing process and not taking account of the heat given off by the reaction, particularly in combustion. This approach is justified in the case of dilute gaseous mixtures.

In order to study turbulent combustion, we are led to use Favre averages. The balance equation [A3.33], established in the case of an incompressible flow, is therefore no longer valid and will be replaced with another.

A3.3.2. *Remarks about averages and scales*

Statistical treatments of turbulence are generally more sophisticated than is suggested by the presentation in the previous section. It is important to specify the conditions in which we can use a certain type of average. Similarly, it is necessary to define reference scales.

A3.3.2.1. *Types of averages*

The statistical average of a random function $f(\vec{\mathbf{x}},t)$ is obtained on the basis of N realizations $f_i(\vec{\mathbf{x}},t)$, such that

$$\langle f\rangle(\vec{\mathbf{x}},t)=\lim_{N\to\infty}\left(\frac{1}{N}\sum_{i=1}^{N}f_i(\vec{\mathbf{x}},t)\right) \quad \text{[A3.48]}$$

More generally, we introduce a probability density $P(f|\vec{\mathbf{x}},t)$ and the average becomes

$$\langle f\rangle(\vec{\mathbf{x}},t)=\int_{-\infty}^{+\infty}f(\vec{\mathbf{x}},t)P(f|\vec{\mathbf{x}},t)\,df \quad \text{[A3.49]}$$

This is the *ensemble average*. Also, for random couples of variables, we can use the density $P(f|\vec{\mathbf{x}},t|\vec{\mathbf{x}}',t')$ which gives us access to the correlations.

When the probability $P(f|\vec{\mathbf{x}},t)$ at a point is not explicitly time-dependent, we are experiencing steady turbulence. When it is independent of the spatial variables, we find ourselves in a case of *homogeneous turbulence*.

We are dealing with *isotropic turbulence* in the case of invariance in relation to the rotations.

Experimentally, we only have a realization at variable times or positions. Therefore, we have to use *spatio-temporal averages* rather than ensemble averages. Denoting the weighting function as $\phi(\vec{\mathbf{x}},t)$, we write

$$\langle f\rangle(\vec{\mathbf{x}},t)=\int_{\vec{\mathbf{x}},t} f\left(\vec{\xi},t\right)\phi\left(\vec{\mathbf{x}}-\vec{\xi},t-\tau\right)d\vec{\xi}\,d\tau \qquad \text{[A3.50]}$$

where $\int_{\vec{\mathbf{x}},t} \phi\left(\vec{\mathbf{x}}-\vec{\xi},t-\tau\right)d\vec{\xi}\,d\tau=1$.

The equivalence of the ensemble averages and spatio-temporal averages requires the *hypothesis of ergodism* to be satisfied.

Also, for a couple of random variables, we introduce spatio-temporal averages at two points.

A3.3.2.2. *Taylor hypothesis*

The *Taylor hypothesis* is valid only in the case of low degrees of turbulence $\overline{v_i'v_j'}^{1/2} << \left(\overline{v_i}\,\overline{v_j}\right)^{1/2}$. It involves supposing that, in the homogeneous case, the turbulent field is at rest and the fluctuations are transported by the motion with average velocity $\overline{\vec{\mathbf{v}}}$ supposed to be constant locally. Thus, we obtain a relation between the spatial and temporal dependencies.[10] With this hypothesis, we need only experimentally measure the fluctuations over time with a fixed probe. The spatial fluctuations will be deduced from this by writing that $f(t)=f(x/\overline{u})$ and $d/dt \cong \overline{u}\,\partial/\partial x$.

A3.3.2.3. *Scales of turbulence*

Integral scale

The *integral scale* ℓ characterizes the large structures in the turbulence. We shall see in section A3.3.4 that the turbulent kinetic energy is distributed on the basis of the wavenumber. The spectral distribution, obtained by way of a Fourier

10 We therefore accept a single rate of advection, which is not possible in the case of compressible fluids (as in acoustics), or for sheared flows where $\overline{u}$ is dependent on the scale.

transform in the homogeneous case, is $E(k)$ per unit mass and wavenumber. We can see that the integral scale is equal to:

$$\ell = \frac{\pi}{2}\frac{E(0)}{\overline{u'^2}} \qquad \text{[A3.51]}$$

$E(k)$ being the one-dimensional spectral distribution.

Kolmogorov scale

The scale of the small structures can be defined as the size beyond which the structures are strongly influenced by viscosity (dissipative region). Thus, we obtain the *Kolmogorov micro-scale* ℓ_K discussed in section A3.3.4.

Taylor micro-scale

The *Taylor micro-scale* λ is an intermediary scale between the integral and Kolmogorov scales. It can be used to link the variances in the slopes of the fluctuations in velocity to the variances of the fluctuations in velocity, and we can define it as:

$$\lambda = \left[\overline{u'^2} \Big/ \overline{\left(\partial u'/\partial x\right)^2}\right]^{1/2} \qquad \text{[A3.52]}$$

We can show, in spectral analysis, that it can be deduced from the energy spectrum $E(k)$: $\lambda^2 \int_0^\infty k^2 E(k)\,dk = \int_0^\infty E(k)\,dk$.

A3.3.3. $k-\varepsilon$ *models (closure for transfer terms)*

Here, we shall only examine the case of a non-reactive fluid, but there are similar treatments that can be performed when dealing with chemical kinetics. Remember that the general balance equation [A3.32] becomes [A3.33], passing through the average, in the incompressible case.

In the case of a compressible fluid, using the Favre averages instead of equation [A3.33], we would have:

$$\partial\left(\bar{\rho}\tilde{f}\right)/\partial t + \vec{\nabla}\cdot\left(\bar{\rho}\widetilde{\tilde{f}\,\vec{\mathbf{v}}}\right) + \vec{\nabla}\cdot\overline{\vec{J}_F} + \vec{\nabla}\cdot\left(\bar{\rho}\widetilde{f''\vec{\mathbf{v}}''}\right) = \overline{\dot{W}_F} \qquad \text{[A3.53]}$$

A3.3.3.1. *$k-\varepsilon$ models in an incompressible fluid*

The averaged mass- and momentum balance equations for a non-reactive incompressible fluid, with constant kinematic viscosity, are written as:

$$\begin{cases} \vec{\nabla}\cdot\overline{\vec{\mathbf{v}}}=0 \\ \partial\overline{\vec{\mathbf{v}}}/\partial t+\vec{\nabla}\cdot\left(\overline{\vec{\mathbf{v}}}\otimes\overline{\vec{\mathbf{v}}}\right)+\vec{\nabla}\left(p/\rho\right)-\vec{\nabla}\cdot\left(2\nu\overline{\vec{\vec{\mathbf{D}}}}\right)+\vec{\nabla}\cdot\overline{\vec{\mathbf{v}}'\otimes\vec{\mathbf{v}}'}=\vec{\mathbf{0}} \end{cases} \qquad [A3.54]$$

Closure of the system is obtained if we know the turbulent transfer coefficient μ_t and the average kinetic energy for the turbulence. Indeed, more accurately than in the previous paragraph, we shall write $\overline{\vec{\mathbf{v}}'\otimes\vec{\mathbf{v}}'}-(1/3)\overline{v'^2}\,\vec{\vec{\mathbf{1}}}=-2\nu_t\,\overline{\vec{\vec{\mathbf{D}}}}$, in the case of isotropic turbulence, the tensor on the left-hand side being the turbulent stresses tensor, with a zero trace. The coefficient $\nu_t=\mu_t/\rho$ can be evaluated on the basis of the Prandtl mixing length L_P. However, a more sophisticated theory is able to link the average kinetic energy of the turbulence per unit mass $\overline{k}=\overline{v'^2}/2$ and to the average dissipation of the turbulence: $\overline{\varepsilon}=\nu\overline{\left(\vec{\nabla}\otimes\vec{\mathbf{v}}'+\left(\vec{\nabla}\otimes\vec{\mathbf{v}}'\right)^T\right):\vec{\nabla}\otimes\vec{\mathbf{v}}'}$. Therefore, we have $\nu_t=\overline{v'l_u'}\cong a_v L\sqrt{\overline{k}}$, where a_v is a constant coefficient and L a mixing length, which might be different to L_P.

We can show that with large Reynolds numbers, the rate of dissipation is linked to L and $\overline{k}$ by the relation: $\overline{\varepsilon}=c_1\overline{k}^2/L$. It follows that:

$$\nu_t=c_v\,\overline{k}^2/\overline{\varepsilon} \qquad [A3.55]$$

where c_v is constant.

On the basis of the local mass- and momentum balance equations, and thanks to adequate multiplications by v'_j and its derivatives and passage through the average, we obtain two balance equations in $\overline{k}$ and $\overline{\varepsilon}$, which bring into play a series of 2nd- or 3rd-order correlation terms and their derivatives. Closure of the system is obtained by neglecting some of these terms and making approximations for the others. The result, which will not be demonstrated here, is as follows:

$$\begin{cases} \partial\overline{k}/\partial t+\overline{\vec{\mathbf{v}}}\cdot\vec{\nabla}\overline{k}=\vec{\nabla}\cdot\left(\nu_t\vec{\nabla}\overline{k}/\sigma_k\right)+\overline{P}(k)-\overline{\varepsilon} \\ \partial\overline{\varepsilon}/\partial t+\overline{\vec{\mathbf{v}}}\cdot\vec{\nabla}\overline{\varepsilon}=\vec{\nabla}\cdot\left(\nu_t\vec{\nabla}\overline{\varepsilon}/\sigma_\varepsilon\right)+c_{\varepsilon 1}\overline{P}(k)-c_{\varepsilon 2}\overline{\varepsilon}^2/\rho\overline{k} \end{cases} \qquad [A3.56]$$

On the right-hand sides of these equations, the first terms are terms of flux, σ_k is a turbulent Prandtl number of energy, σ_ε a turbulent Prandtl number of dissipation. The other terms are terms of production and destruction. We have $\bar{P}(k) = -\nu_t \overline{\vec{\mathbf{v}}'\otimes\vec{\mathbf{v}}'} : \vec{\nabla}\otimes\bar{\vec{\mathbf{v}}}$.

The constants have been adjusted over numerous experiments, and a judicious choice seems to be:

$$c_\mu = 0.09,\ \sigma_k = 1,\ \sigma_\varepsilon = 1.3,\ c_{\varepsilon 1} = 1.44,\ c_{\varepsilon 2} = 1.92.$$

The equations obtained no longer involve correlations but simply the slopes of the averaged values. They are coupled with the averaged balance equations, with the whole of the system being solvable by numerical methods, with the appropriate choice of initial and boundary conditions.

A3.3.3.2. *$k-\varepsilon$ models in compressible flow*

By neglecting the terms of flotation and volume viscosity, the averaged Navier–Stokes equations can be written thus:

$$\begin{cases} \partial\bar{\rho}/\partial t + \vec{\nabla}\cdot\left(\bar{\rho}\tilde{\vec{\mathbf{v}}}\right) = 0 \\ \partial\left(\bar{\rho}\tilde{\vec{\mathbf{v}}}\right)/\partial t + \vec{\nabla}\cdot\left(\bar{\rho}\tilde{\vec{\mathbf{v}}}\otimes\tilde{\vec{\mathbf{v}}}\right) + \vec{\nabla}\bar{p} - \vec{\nabla}\cdot\left(2\bar{\rho}\nu\tilde{\bar{\bar{\mathbf{D}}}}\right) + \vec{\nabla}\cdot\left(\bar{\rho}\widetilde{\vec{\mathbf{v}}''\otimes\vec{\mathbf{v}}''}\right) = \vec{\mathbf{0}} \end{cases} \quad \text{[A3.57]}$$

This time, we have $\bar{\rho}\widetilde{\vec{\mathbf{v}}''\otimes\vec{\mathbf{v}}''} = -\bar{\rho}\nu_t\left(2\tilde{\bar{\bar{\mathbf{D}}}} - (2/3)\vec{\nabla}\cdot\tilde{\vec{\mathbf{v}}}\,\bar{\bar{\mathbf{1}}}\right) + (2/3)\bar{\rho}\tilde{k}\,\bar{\bar{\mathbf{1}}}$ and $\tilde{k} = (1/2)\widetilde{\vec{\mathbf{v}}''\cdot\vec{\mathbf{v}}''}, \nu_t = c_\mu \bar{k}^2/\varepsilon, c_\mu = 0.09$. The k-ε system of equations is then as follows:

$$\begin{cases} \bar{\rho}\partial\tilde{k}/\partial t + \bar{\rho}\tilde{\vec{\mathbf{v}}}\cdot\vec{\nabla}\tilde{k} = \vec{\nabla}\cdot\left(\bar{\rho}\nu_t\vec{\nabla}\tilde{k}/\sigma_k\right) + \tilde{P}(k) - \bar{\rho}\tilde{\varepsilon} \\ \bar{\rho}\partial\tilde{\varepsilon}/\partial t + \bar{\rho}\tilde{\vec{\mathbf{v}}}\cdot\vec{\nabla}\tilde{\varepsilon} = \vec{\nabla}\cdot\left(\bar{\rho}\nu_t\vec{\nabla}\tilde{\varepsilon}/\sigma_\varepsilon\right) + c_{\varepsilon 1}\left(\bar{\rho}\tilde{\varepsilon}/\tilde{k}\right)\tilde{P}(k) - c_{\varepsilon 2}\bar{\rho}\tilde{\varepsilon}^2/\tilde{k} \end{cases} \quad \text{[A3.58]}$$

with the same coefficients as in section A3.3.3.1 and $\tilde{P}(k) = -\bar{\rho}\nu_t\widetilde{\vec{\mathbf{v}}''\otimes\vec{\mathbf{v}}''} : \vec{\nabla}\otimes\tilde{\vec{\mathbf{v}}}$.

A3.3.3.3. *Remarks about $k-\varepsilon$ models*

There are various criticisms that could be leveled at "$k-\varepsilon$"-type models. They give an irrelevant solution in the homogeneous and isotropic case ($\bar{\vec{\mathbf{v}}} = \vec{\mathbf{0}}$ in an incompressible case). They are based on a one-point formalism and therefore do not contain all the spectral information [AND 80; LES 97].

It is noteworthy that the $k-\varepsilon$ simulation is not compatible with the energy degradation law in spectral theory. Indeed, in the case of isotropic turbulence, the $k-\varepsilon$ model is reduced to $\partial k/\partial t=-\overline{\varepsilon};\partial\overline{\varepsilon}/\partial t=-c_{\varepsilon 2}\overline{\varepsilon}^2/k$, which leads us to $\partial^2 k/\partial t^2=-c_{\varepsilon 2}\left(\partial k/\partial t\right)^2/k$ or, after integration, $k=\left(At+B\right)^{-1/(c_{\varepsilon 2}-1)}$. Yet in the $k-\varepsilon$ model, the energy decreases in $t^{-\alpha_\varepsilon}$, where α_ε is independent of the form of the initial spectrum, unlike what is observed in the real world.

Nevertheless, $k-\varepsilon$ models are frequently used, including in reactive flows, where the problem also arises of how to determine the average terms of chemical production $\overline{\dot{W}_j}$.

A3.3.4. *Spectral analysis and Kolmogorov's theory*

The Fourier transform $f\left(\vec{\mathbf{k}}\right)=\frac{1}{8\pi^3}\int_{\Re^3} f\left(\vec{\mathbf{x}}\right)e^{-i\vec{\mathbf{k}}\cdot\vec{\mathbf{x}}}d\vec{\mathbf{x}}$ enables us to study the energy distribution of the spatial structures of the turbulence. The inverse of the wavenumber *k* defines the scale of the turbulence, i.e. the size of the vortices.

We shall not go into detail about spectral theory – in fact we shall only briefly touch upon it here. Note, however, that today energy spectra can be obtained fairly easily experimentally. Here, we shall assume that the turbulence is isotropic and therefore homogeneous ($\langle\vec{\mathbf{v}}\rangle=\vec{\mathbf{0}}$ in particular) and that the flow is incompressible and viscous. In addition, the evolution is free and we ignore the influence of the walls. We can then show that the turbulent energy spectrum *E(k)* includes two zones separated by a transition zone (Figure A3.13):

– a zone with small values of *k*, i.e. a zone with large eddies (size k_e^{-1}) where the energy is concentrated and that can be steadily fed with energy (by instability, interaction with the average motion, external forces, etc.) at a rate $\dot{e}_1$;

– a zone of small structures of size $1/k_d$ (k_d Kolmogorov wavenumber) where the energy is dissipated by viscosity at a rate $\dot{e}_2$;

– a direct cascade where the advection terms (in $\partial\left(u_i u_j\right)/\partial x_j$) take care of the conservative transfer of energy between the scales.

There is also an inverse cascade, of lesser energetic significance, which corresponds to the grouping of small eddies and therefore to the transfer of some of the turbulent energy to the larger structures.[11] This energy transfer phenomenon is predominant when the wavenumbers are very low.

In order for the transfer zone to have a non-negligible extent, it is necessary for $k_d >> k_e$; this implies a condition on the Reynolds numbers. We define a Reynolds number for the turbulence on the basis of the size k^{-1} of the vortices and a velocity based on the energy *E(k)* (which is an energy of fluctuations in velocity per unit mass and per unit wavenumber). The value $\left[k E(k)\right]^{1/2}$ is therefore the average velocity that is characteristic of structures of dimension k^{-1}, and the associated Reynolds number is therefore $\mathrm{Re} = (1/\nu k)\left[k E(k)\right]^{1/2}$.

For $k = k_d$, the dissipative phenomena become as significant as the inertial phenomena, so that $(\mathrm{Re})_d = (1/\nu k_d)\left[k_d E(k_d)\right]^{1/2}$.

For large structures, we obtain $(\mathrm{Re})_e = (1/\nu k_e)\left[k_e E(k_e)\right]^{1/2}$ where $E(k_e) >> E(k_d)$ and $k_e << k_d$. It follows from this that $(\mathrm{Re})_e >> (\mathrm{Re})_d \cong 1$.

The inequality summarizes the conditions of existence of a transfer zone where the motions are independent of the large energy scales, sources of turbulence and of viscosity.

Kolmogorov's theory postulates:

– the existence of small scales;

– the existence of an inertial zone (vortex cascade hypothesis).[12]

It also enables us to determine the spectrum $E(k)$ by way of dimensional analysis (see Appendix 1).

---

11 In spectral analysis, we consider triads $(\mathbf{k},\mathbf{q},\mathbf{p})$ of wavenumber vectors forming a closed triangle (such that $\mathbf{k}+\mathbf{q}+\mathbf{p}=\mathbf{0}$). With a direct cascade $k_d >> k$ (non-local cascade), we see drainage of energies from large scales to small scales (from $\mathbf{k}$ to the larger vectors $\mathbf{q}$ and $\mathbf{p}$). With an inverse cascade, the vector $\mathbf{k}$ is larger than at least one of the vectors $\mathbf{q}$ and $\mathbf{p}$ and therefore feeds larger structures [SAG 98].

12 It is possible to define a discrete sequence of vortical entities in the inertial zone of turbulence, with $l_k = \ell/2^n \geq l_k$, as done by Peters [PET 00].

We suppose the regime to be quasi-steady; therefore the rates $\dot{e}_1$ and $\dot{e}_2$ have a common value, which is an energy produced per unit mass in the unit time (dimension L^2T^{-3}). We can directly connect ε to the speed of rotation and to the integral scale ℓ of the large eddies: $\varepsilon \approx v'^3/\ell$. Similarly, at the speed of rotation v_n of the eddies in each class of inertia with size ℓ_n, then $\varepsilon \approx v_n^3/\ell_n$, where the rate ε is the same.

We can take $k, \bar{\varepsilon}, \nu, E(k)$ as characteristic parameters of the problem. It is common to replace ν with a combination between $\bar{\varepsilon}$ and ν, with the dimensions of a length: the Kolmogorov scale defined in section A3.3.2.3[13]:

$$\ell_k = \left(\nu^3/\bar{\varepsilon}\right)^{1/4} \qquad \text{[A3.59]}$$

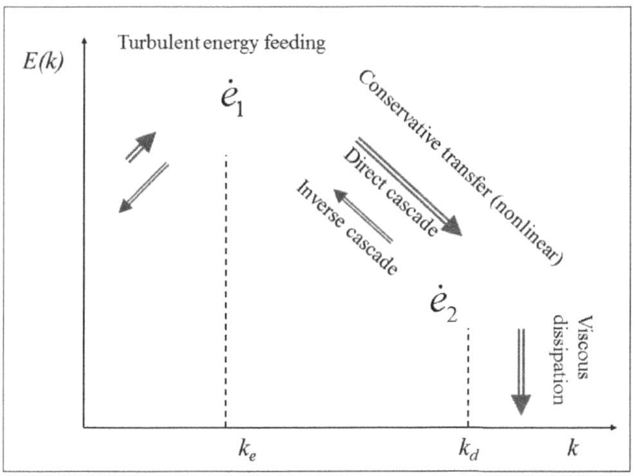

Figure A3.13. *Characteristic zones of the turbulent energy spectrum*

We then choose k and $\bar{\varepsilon}$ as basic values. The Π groups can then be deduced from Table A3.2:

$$\Pi_l = k\ell_k, \; \Pi_E = k^{5/3}\bar{\varepsilon}^{-2/3}E(k) \qquad \text{[A3.60]}$$

---

13 The size and velocity of the Kolmogorov vortices obey the relation $\varepsilon = v_k^3/\ell_k$.

	k	$\overline{\varepsilon}$	ℓ_k	$E(k)$
L	-1	2	1	3
M	0	0	0	0
T	0	-3	0	-2

Table A3.2. *Dimensions of the quantities in the problem*

The Vaschy–Buckingham theorem gives us $\Pi_E = \psi(\Pi_l)$, or indeed

$$E(k) = \overline{\varepsilon}^{2/3} k^{-5/3} \Psi(k\ell_k) \quad \text{[A3.61]}$$

In the inertial zone, viscosity has no effect and we have $\Psi(k\ell_k) = c_k$, the Kolmogorov constant, and

$$E(k) = c_k \, \overline{\varepsilon}^{2/3} k^{-5/3} \quad \text{[A3.62]}$$

Experience has validated this law for non-reactive media.[14] The constant c_k is of the order 1.5.

We can also show that at very small values of *k*, less than k_e, the energy spectrum is in k^s where $s = 4$. This remark is also applicable to large-eddy simulation [SAG 98].

Figure A3.14 relates to the one-dimensional spectrum, but the law is a similar one in three dimensions, with a constant $48c_k/55$.

The inertial zone is limited by the wavenumber k_d which, as $\text{Re}_d \approx 1$, is equal to $k_d \approx E(k_d)/\rho v^2$. In the inertial zone, for $k = k_d$, we therefore have $k_d \approx \overline{\varepsilon}^{2/3} k^{-5/3}/v^2$. It follows that in the dissipative zone:

$$E(k) = \overline{\varepsilon}^{2/3} k^{-5/3} \psi(k/k_d) \quad \text{[A3.63]}$$

14 In the presence of chemical reactions, particularly in combustion, the $k^{-5/3}$ law is no longer verified. This is doubtless due to the fact that the turbulence is then no longer homogeneous or isotropic. In addition, other parameters come into play in addition to those taken into account here – those due to chemical productions, to the diffusion of species and to heat transfer, as well as geometric parameters.

In addition, the Reynolds number for the large eddies is

$$\mathrm{Re}_e = \left(1/\nu k_e\right)\left[k_e E\left(k_e\right)\right]^{1/2} \Rightarrow \mathrm{Re}_e \approx \left(k_e^{-3/4}/\nu\right)\bar{\varepsilon}^{1/3} = \left(k_d/k_e\right)^{4/3} >> 1$$

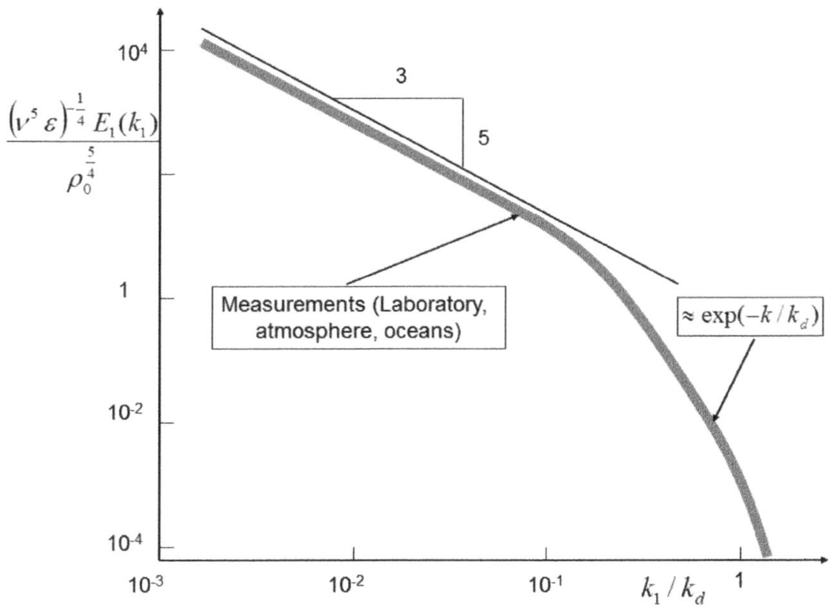

Figure A3.14. *One-dimensional energy spectrum illustrating the universality of Kolmogorov's law*

A3.4. Ideas about large eddy simulation

Direct numerical simulation (DNS), a method which consists of numerically solving the balance equations for the fluxes, is currently the most accurate method. It is particularly appropriate for flows with a small Reynolds number. However, it becomes prohibitive with large Reynolds numbers, because the levels of power of modern computers are insufficient (even when parallel computing is used).

The difficulties encountered with statistical methods and the inability to use DNS for flows with large Reynolds numbers has led to the use, since the 1960s, of a hybrid method: Large Eddy Simulation (LES) [SMA 63].

The objective of LES is to explicitly compute the larger turbulent structures in the field of flow (typically structures which are large than the computation mesh size), while the smaller effects are modeled.

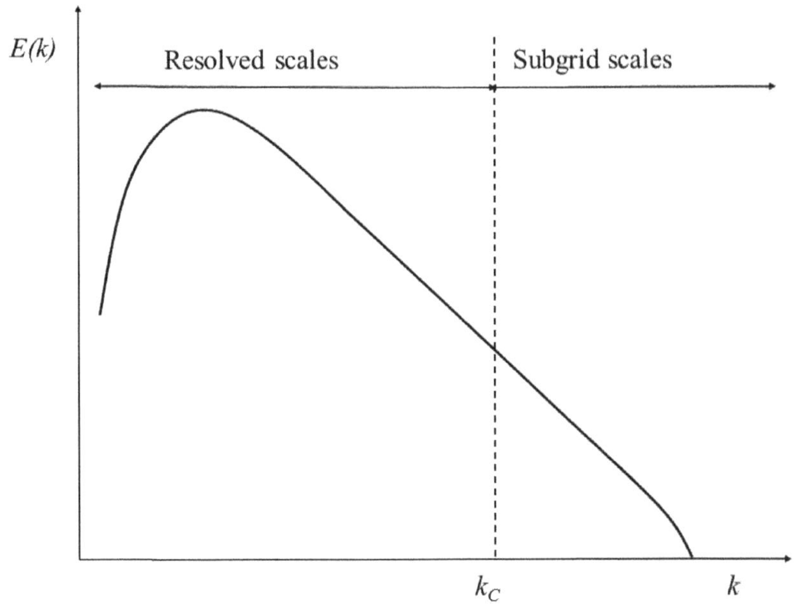

Figure A3.15. *Cut-off wavenumber*

Our aim here is not to give a complete overview of this method, which is used in many modeling studies, because to do so would go beyond the remit of this book. We shall limit our examination to incompressible flows. Readers can refer to other works for further details on flows of incompressible simple fluids [LES 97; SAG 01]. A few brief guidelines about the use of LES in combustion are given in Chapter 3.

A3.4.1. *Filtering*

In LESs, the variables are filtered in the spectral space or in the physical space. In the spectral space, we consider a cut-off wavenumber k_c of the spectrum *E (k)* of turbulent kinetic energy (Figure A3.15).

The objective here is to give as full a treatment as possible of small eddies in the turbulence (*or unresolved scales or "subgrid" scales*, or indeed eddies *modeled in LESs defined by the inequality* $k > k_c$) statistically. We then need to establish valid equations for large eddies (*or resolved scales, or computed by LESs*) and solve these equations directly.

We need to take account of the interactions between the modes at different scales and attempt to reintroduce the concept of turbulent viscosity. The first task is to establish a system of equations which need to be solved for the different eddies and are based on classic balance laws.

We call $G = G(\vec{\mathbf{x}},t)$ a filter in the physical space. When it is applied to any quantity $f(\vec{\mathbf{x}},t)$, this filter gives us the resolved part $\bar{f}(\vec{\mathbf{x}},t)$ [15]:

$$\bar{f}(\vec{\mathbf{x}},t) = \int_{\Re^3} f(\vec{\mathbf{x}},t) G\left(\vec{\mathbf{x}} - \vec{\xi}, t - t'\right) dt' d\vec{\xi} \qquad \text{[A3.64]}$$

or, in symbolic terms, $\bar{f} = Gf$. Conversely, the non-resolved part will be $f' = (1-G)f$. Transposed into the Fourier space, these definitions give us $\bar{f} = G(k)f, f' = (1 - G(k))f$, where this time, $G(k) = G(\vec{\mathbf{k}},\omega), f = f(\vec{\mathbf{k}},\omega)$.

We shall now present three filters that are commonly used in LESs (the forms of these filters in the physical space and in the spectral space are shown in Figure A3.16):

– in the spectral space, the *sharp cut-off filter* is:

$$G(k) = 1, \left|\vec{\mathbf{k}}\right| \le k_c, \quad G(k) = 0, \left|\vec{\mathbf{k}}\right| > k_c$$

with $k_c = \pi/\Delta$ (Δ is the size of the filter; the filter keeps back eddies of length greater than 2Δ). In the physical space, the corresponding filter is $G(x) = \sin(k_c x)/k_c x$;

– a *box filter* (or top-hat filter) has $G(x) = 1/\Delta$ for $|x| \le \Delta/2$ and 0 elsewhere. It becomes $\hat{G}(k) = \sin(k\Delta/2)/(k\Delta/2)$ in the spectral space. In the present case, this criterion corresponds to an average over a box with dimensions Δ;

---

15 In the absence of ambiguity, we shall preserve the same notations for the filtered quantities in equation [A3.63] and for the classic average values defined by [A3.25]. This will also remain valid for the filtered Favre values defined by [A3.30].

– a *Gaussian filter* $G(x)=\sqrt{\gamma/\pi\Delta^2}\exp(\gamma x^2/\Delta^2)$ in the physical space, where γ is a constant, gives another Gaussian filter $G(k)=\exp(-\Delta^2 k^2/4\gamma)$ in the spectral space.

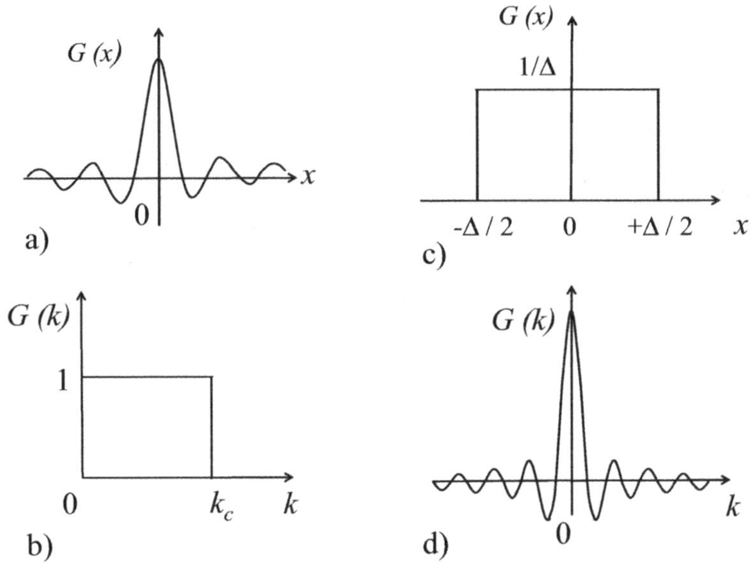

Figure A3.16. *Examples of filters (from [SAG 01]): a) cut-off filter in the physical space; b) cut-off filter in the spectral space; c) box filter in the physical space; d) box filter in the spectral space*

All these filters are normalized: $\int_{-\infty}^{+\infty} G(x)\,dx=1$ in the 1D case. The Favre average can also be defined for turbulent flows with variable density by way of:

$$\bar{\rho}\tilde{f}(\vec{x},t)=\int_{\Re^3}(\rho f)(\vec{x},t)G(\vec{x}-\vec{\xi},t-t')dt'd\vec{\xi} \qquad \text{[A3.65]}$$

We shall now follow in the footsteps of Ghosal and Moin [GHO 95], and list the properties which indicate that a filter is acceptable in the one-dimensional homogeneous case:

– it is symmetrical;

– the constants are preserved;

– it decreases rapidly;

– it exhibits quasi-locality in the physical space.

Note also that [POI 05]:

– unlike with the RANS average, the filtered value of a LES disturbance is non-null: $\overline{f'} \neq 0$. Thus, the filtered and double-filtered values are usually not equal: $\overline{\overline{f}} \neq \overline{f}$. Similarly, $f" = f - \tilde{f}, \widetilde{f"} \neq 0$ and $\tilde{\tilde{f}} \neq \tilde{f}$;

– to establish the balance equations for the filtered values, it is necessary to exchange the filtering and differentiation operators.

A3.4.2. *Filtered balance equations for a non-reactive incompressible fluid*

After filtration, the incompressible Navier–Stokes equations for a non-reactive fluid with constant viscosity become [GHO 95]:

$$\begin{cases} \dfrac{\partial \overline{u}_i}{\partial x_i} = 0 \\ \dfrac{\partial \overline{u}_i}{\partial t} + \dfrac{\partial \left(\overline{u_i u_j} \right)}{\partial x_j} = -\dfrac{\partial \overline{p}}{\partial x_i} + \nu \dfrac{\partial}{\partial x_j} \left(\dfrac{\partial \overline{u}_i}{\partial x_j} + \dfrac{\partial \overline{u}_j}{\partial x_i} \right) \end{cases} \quad [A3.66]$$

Unlike in equation [A3.33], the $\overline{f}$ values this time are averages of filtered values.

The filtered balance equations which are usable in cases of combustion are given in Chapter 3.

A3.4.3. *Closure relations for the filtered balance equations*

We are now going to summarize the main approaches for unresolved modeling of the transport terms (unresolved Reynolds tensions tensor: $\vec{\vec{\mathcal{T}}} = \widetilde{\vec{v} \otimes \vec{v}} - \tilde{\vec{v}} \otimes \tilde{\vec{v}}$, unresolved scalar fluxes). The case of a reactive medium is discussed in Chapter 3, with the filtered laminar diffusion flux, and the filtered chemical reaction rates. The models below can only be validated by comparison with DNS and the experimental results (where possible).

The *unresolved Reynolds tension* models were first presented in [POI 05], where the density was taken to be constant for simplicity's sake.

Here, we shall use the tensorial symbols and notations (see section A1, [PRU 12]).

A3.4.3.1. *Smagorinsky model*

Just like with conventional averages, turbulent fluxes are supposed to be proportional to turbulent forces. This leads to the introduction of a turbulent viscosity of subgrid scales, which is different to the turbulent viscosity defined by [A3.54]. We write

$$\vec{\vec{\mathcal{T}}} - \frac{1}{3}\mathrm{tr}\vec{\vec{\mathcal{T}}}\,\vec{\vec{\mathbf{1}}} = -\nu_t \vec{\nabla} \otimes \bar{\vec{\mathbf{v}}} = -2\nu_t \bar{\vec{\vec{\mathbf{S}}}} \qquad \text{[A3.67]}$$

where $\bar{\vec{\vec{\mathbf{S}}}}$ is the symmetrical part of the tensor $\vec{\nabla} \otimes \bar{\vec{\mathbf{v}}}$, and we have $\nu_t = C_S{}^2 \Delta^{4/3} \ell^{2/3} \left|\bar{\vec{\vec{\mathbf{S}}}}\right|$, where ℓ is the integral scale of the turbulence, C_s is a constant, and $\bar{S} = \left(2\bar{\vec{\vec{\mathbf{S}}}} : \bar{\vec{\vec{\mathbf{S}}}}\right)^{1/2}$ is the resolved strain rate. The expression of ν_t can be simplified by assuming that the integral scale is approximately the same as the grid size – that is to say that $\ell \approx \Delta$.

A3.4.3.2. *Scale similarity model*

The scale similarity model is based on double filtering and on the idea that the unresolved stresses are primarily controlled by the larger unresolved structures (see Figure A3.17a), which are similar to the smaller resolved structures. The subgrid tensor is then evaluated using

$$\vec{\vec{\mathcal{T}}} = \overline{\bar{\vec{\mathbf{v}}} \otimes \bar{\vec{\mathbf{v}}}} - \bar{\bar{\vec{\mathbf{v}}}} \otimes \bar{\bar{\vec{\mathbf{v}}}} \qquad \text{[A3.68]}$$

As this closure is not sufficiently dissipative, it is usually combined with a turbulent viscosity model such as the Smagorinsky model, giving us a hybrid model [LES 97].

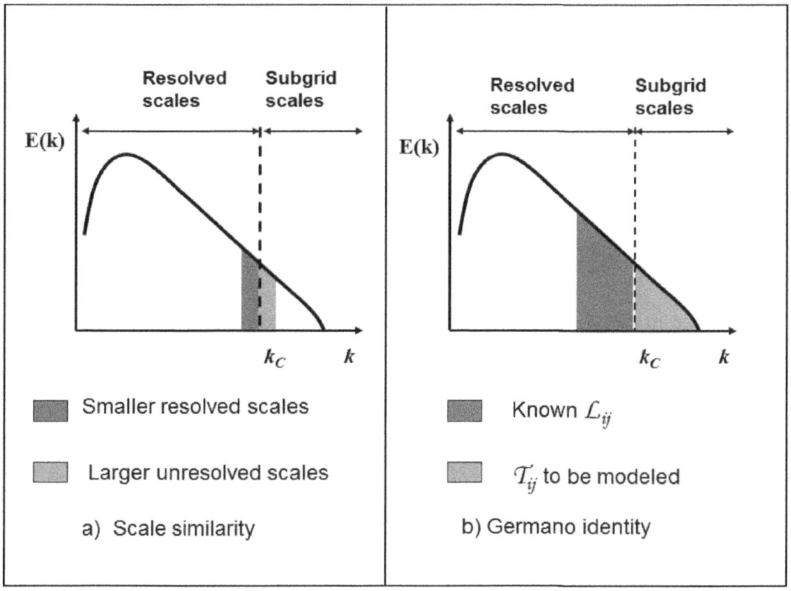

Figure A3.17. *Scale similarity and Germano dynamic model*

A3.4.3.3. *Germano dynamic model*

A test filter $\hat{\Delta}$, which is larger than the LES filter (of dimension $\overline{\Delta}$) is now introduced [GER 91] (see Figure A3.17b). For unresolved momentum fluxes, if we remove the arrows from the 1st- and 2nd-order tensors for ease of writing, we have

$$\mathcal{T} = \overline{\mathbf{v}\otimes\mathbf{v}} - \overline{\mathbf{v}}\otimes\overline{\mathbf{v}} \qquad [A3.69]$$

and for unresolved fluxes at the test level, we have

$$\hat{\mathcal{T}} = \widehat{\overline{\mathbf{v}\otimes\mathbf{v}}} - \widehat{\overline{\mathbf{v}}\otimes\overline{\mathbf{v}}} \qquad [A3.70]$$

We then introduce the "bar hat" double filter:

$$\mathbf{T} = \widehat{\overline{\mathbf{v} \otimes \mathbf{v}}} - \hat{\bar{\mathbf{v}}} \otimes \hat{\bar{\mathbf{v}}} \qquad \text{[A3.71]}$$

Now, we consider the resolved tensor corresponding to the test filter applied to the field $\bar{\mathbf{v}}$:

$$\mathcal{L} = \widehat{\bar{\mathbf{v}} \otimes \bar{\mathbf{v}}} - \hat{\bar{\mathbf{v}}} \otimes \hat{\bar{\mathbf{v}}} \qquad \text{[A3.72]}$$

By adding equations [A3.69] and [A3.70] together we obtain

$$\mathcal{L} = \mathbf{T} - \hat{\mathcal{T}} \qquad \text{[A3.73]}$$

Equation [A3.72] is called the *Germano identity*. The Reynolds tensors $\mathcal{T}$ and T can be estimated on the basis of the Smagorinsky model (equation [A3.66] with $\ell = \Delta$):

$$\begin{cases} \mathcal{T} - \dfrac{1}{3}\mathrm{tr}\mathcal{T}\mathbf{1} = -2C\bar{\Delta}^2 \left|\bar{\mathbf{S}}\right| \bar{\mathbf{S}} - 2C\boldsymbol{\alpha} \\ \hat{\mathcal{L}} = \mathbf{T} - \hat{\mathcal{T}} = -2C\bar{\Delta}^2 \left|\hat{\bar{\mathbf{S}}}\right| \hat{\bar{\mathbf{S}}} = -2C\boldsymbol{\beta} \end{cases} \qquad \text{[A3.74]}$$

where C is a parameter which needs to be determined. The Germano identity (equation [A3.72]) is then rewritten as:

$$\mathcal{L} - \frac{1}{3}\mathrm{tr}\mathcal{L}\mathbf{1} = 2C\left(\hat{\boldsymbol{\alpha}} - \boldsymbol{\beta}\right) \qquad \text{[A3.75]}$$

In order to determine $C = C(\mathbf{x}, t)$, we also have, in the incompressible case, the two mass conservation equations $\mathrm{tr}\left(\bar{\mathbf{S}}\right) = \mathrm{tr}\left(\hat{\bar{\mathbf{S}}}\right) = 0$. By calculating C, we get instability, and a least-squares approach is used to minimize the error.

This model has also been extended to compressible turbulence [MOI 91]. It is a highly effective model for numerous applications.

A3.4.3.4. *Structure function model*

These models are based on a theoretical analysis of turbulence in the spectral space and on the concept of subgrid viscosity. The dynamic subgrid viscosity is

$$\nu_t(\bar{\mathbf{x}},\Delta)=0.105c_K^{-3/2}\Delta\sqrt{F_2(\bar{\mathbf{x}},\Delta)} \qquad \text{[A3.76]}$$

where $c_K = 1.4$ is the Kolmogorov constant (see equation [A3.61]) and F_2 is the structure function $F_2(\mathbf{x},\Delta)=\left[\bar{\mathbf{v}}(\mathbf{x}+\mathbf{r})-\bar{\mathbf{v}}(\mathbf{x})\right]^2$, when $\Delta=|\mathbf{r}|$. A high-pass filter is used to eliminate the large eddies (which cause major dissipation) with the filtered structure function $\overline{F_2}$. The resulting subgrid viscosity is then:

$$\nu_t(\bar{\mathbf{x}},\Delta)=0.105c_K^{-3/2}\Delta\sqrt{F_2(\bar{\mathbf{x}},\Delta)} \qquad \text{[A3.77]}$$

A3.5. Conclusion

The elements presented in this section illustrate the complexity of the topic of turbulence. Fluid turbulence is currently considered to be one of the most difficult problems to resolve [ORL 01; SAG 06]. "One does not know, for instance, how to predict the critical Reynolds number of transition to turbulence in a pipe, nor how to compute precisely the drag of a car or an aircraft, even with today's largest computers" (see the foreword by M. Lesieur in [SAG 01]).

We have merely presented some of the methods used in this domain: spectral analysis, $k-\varepsilon$ modeling, and LES. There are other techniques – particularly probability distribution functions or pdf's, which were touched upon in Chapter 2 with regard to residence time distribution (see section 2.4.1) and are used in the context of turbulent combustion in Chapter 3 (section 3.3.4.7) on chemical production rates. Current research often places the emphasis on improvement of the existing models and on comparison of theoretical results with experimental measurements. Other research relates to the behavior of large stable or unstable structures within the flow. Finally, the interaction between the dynamic and chemical effects has not been completely clarified. However, a large proportion of the current knowledge in this domain is put into practice, and although a great deal of progress may remain to be made, we are now basically able to precisely model numerous practical situations.

The transition in the vicinity of the walls is the subject of research with the $k-\varepsilon-f_\mu$ method [PAR 03] and hybrid methods such as OES (*Organized Eddy Simulation*), which combines RANS and LES with moveable walls mobiles

[HAA 09; BRA 12]. The same is true more generally as soon as the turbulence becomes unsteady[16], when we use the URANS methods [MER 09; MAU 13].

16 The hypothesis of steadiness (section A3.3.4) is generally made when studying turbulence. However, it is inappropriate in certain cases when the flow is highly unsteady: vortical detachment in the wake of an obstacle, large structures produced by the instability of a mixture layer, forced fluctuations in the conditions of injection to the opening of a combustion chamber, etc.

Appendix 4

Thermodynamic Functions for a Mixture in Disequilibrium

The mixture under discussion here is air at high temperature without ionization, but with thermal and chemical disequilibria as was the case in section 1.3.

A4.1. Thermodynamics

In line with Dunham's approach as outlined in [MAY 40], the authors of [FLA 93] write the partition function of the species as the product of the partition functions associated with the internal modes, which are functions of their own specific equilibrium temperatures:

$$Q_\alpha = Q_{\alpha;t}\left(V,T\right)Q_{\alpha;r}\left(T\right)Q_{\alpha;el}\left(T\right)Q_{\alpha;v}\left(T_{\alpha;v}\right)Q_{\alpha;r-v}\left(T_{\alpha;v}\right) \qquad \text{[A4.1]}$$

Here, $Q_{\alpha;t}$, $Q_{\alpha;r}$ $Q_{\alpha;el}$ and $Q_{\alpha;v}$ are, respectively, the partition functions of the internal modes of translational, rotational, electronic and vibrational motion of the species. The quantity $Q_{\alpha;r-v}$ represents the partition function of coupling between rotation and vibration, and has been set apart from the others.

The expressions of the partition functions are given:

$$\begin{cases} \ln Q_{\alpha;t} = 3\ln\left(\sqrt{\dfrac{2\pi \mathcal{M}_\alpha k^2}{R h^2}}\right) + \ln V + \dfrac{3}{2}\ln T, \quad \ln Q_{\alpha;r} = \psi_\alpha \ln\left(\dfrac{T}{\theta_{\alpha;r}}\right), \\ \ln Q_{\alpha;el} = \ln\left[\sum_i g_{\alpha;el;i} \exp\left(-\varepsilon_{\alpha;el;i}/kT\right)\right], \quad \ln Q_{\alpha;v} = \psi_\alpha \ln\left[1-\exp\left(\theta_{\alpha;v}/T_{\alpha;v}\right)\right] \\ \ln Q_{\alpha;r-v} = \psi_\alpha \dfrac{\theta_{\alpha;r}}{\theta_{\alpha;v}}\left\{\dfrac{8T_{\alpha;v}}{\theta_{\alpha;v}} + 3\left[\left(\dfrac{x_{e;\alpha}\theta_{\alpha;v}}{\theta_{\alpha;r}}\right)^{1/2} - 1\right]\coth\left(\dfrac{\theta_{\alpha;v}}{2T_{\alpha;v}}\right)\right\} + \psi_\alpha 2x_{e;\alpha}\dfrac{T_{\alpha;v}}{\theta_{\alpha;v}} \end{cases}$$

[A4.2]

with $\psi_\alpha = 1$ for diatomic molecules and $\psi_\alpha = 0$ for monatomic molecules. The volume of the mixture is V, $\varepsilon_{\alpha;el;i}$ is the electron energy for the mode of level i of the species α, $g_{\alpha;el;i}$ the degeneracy of the electronic mode for the level i of the species α, $\theta_{\alpha;v}$ is the characteristic temperature of vibration of the species α, $\theta_{\alpha;r}$ is the characteristic temperature of rotation of the species α and $x_{e;\alpha}$ is the anharmonicity constant of the species α.

The internal energies of the species α are

$$\begin{cases} e_{\alpha;t} = \dfrac{3}{2}\dfrac{R}{\mathcal{M}_\alpha}T, \; e_{\alpha;r} = \psi_\alpha \dfrac{R}{\mathcal{M}_\alpha}T, \\ e_{\alpha;el} = \dfrac{R}{\mathcal{M}_\alpha}\dfrac{\sum_i g_{\alpha;el;i}\left(\varepsilon_{\alpha;el;i}/k\right)\exp\left(-\varepsilon_{\alpha;el;i}/kT\right)}{\sum_i g_{\alpha;el;i}\exp\left(-\varepsilon_{\alpha;el;i}/kT\right)}, \\ e_{\alpha;v} = \psi_\alpha \dfrac{R}{\mathcal{M}_\alpha}\left[\dfrac{\theta_{\alpha;v}}{\exp\left(\theta_{\alpha;v}/T_{\alpha;v}\right)-1} + \left(2x_{e;\alpha} + \dfrac{8\theta_{\alpha;r}}{\theta_{\alpha;v}}\right)\dfrac{T_{\alpha;v}^{\;2}}{\theta_{\alpha;v}}\right] \\ +\psi_\alpha \dfrac{R}{\mathcal{M}_\alpha}\dfrac{3}{2}\theta_{\alpha;r}\left[\left(\dfrac{x_{e;\alpha}\theta_{\alpha;v}}{\theta_{\alpha;r}}\right)^{1/2} - 1\right]\dfrac{1}{\mathrm{sh}^2\left(\theta_{\alpha;v}/2T_{\alpha;v}\right)} \end{cases}$$

[A4.3]

where $e_{\alpha;v}$ contains the harmonic and anharmonic contributions, as well as the rotation/vibration coupling.

The specific heat $c_{v;\alpha;t,r}$ verifies the relations

$$c_{v;\alpha;t,r} = c_{v;\alpha;t} + c_{v;\alpha;r},\ c_{v;\alpha;t} = \frac{3}{2}\frac{R}{\mathcal{M}_\alpha},\ c_{v;\alpha;r} = \psi_\alpha \frac{R}{\mathcal{M}_\alpha} \qquad [A4.4]$$

Here, $c_{v;\alpha;t}$ and $c_{v;\alpha;r}$ are, respectively, the specific heats of translational and rotational motion for the species at constant volume. The different constants are given in Table 1 in [FLA 93].

A4.2. Chemistry

The constants in equation [1.131] are given for each chemical reaction in Table 2 in [FLA 93], and correspond to Gardiner's model [GAR 84].

A4.3. Vibration

The vibrational relaxation times are drawn from the experimental study conducted by Millikan and White [MIL 63]:

$$\begin{cases} 1/\tau_{v;\beta;\gamma} = p_\alpha \exp\left[-B_{\beta;\gamma}\left(T^{-1/3} - C_{\beta;\gamma}\right) + D\right], \\ B_{\beta;\gamma} = 1.16\ 10^{-3}\,\mu_{\beta;\gamma}{}^{1/2}\,\theta_{\beta;v}{}^{4/3},\ C_{\beta;\gamma} = 0.015\,\mu_{\beta;\gamma}{}^{1/4} \\ D = 18.42,\ \mu_{\beta;\gamma} = \dfrac{\mathcal{M}_\beta \mathcal{M}_\gamma}{\mathcal{M}_\beta + \mathcal{M}_\gamma} \end{cases} \qquad [A4.5]$$

where p_α is the pressure of the mixture in atmospheres, and $\mu_{\beta;\gamma}$ is the reduced mass, in grams.

Appendix 5

Notion of Bifurcation

A bifurcation occurs when a small change to a physical parameter λ causes a major change in the organization of a system. This happens, for instance, in section A3.2.2.3, with the Reynolds numbers involved in the changing stability of the flow between two cylinders with different rotations. Other characteristic parameters play a part in other systems.

There are various types of bifurcation which it might be useful to be able to characterize. We commonly find: pitchfork bifurcations, saddle-node bifurcations, Hopf bifurcations and period-doubling bifurcations.

Each type of bifurcation is characterized by a normal form, i.e. by a generally typical equation.

Here, we shall limit our discussion to pitchfork bifurcation.

For instance, the dimensionless Landau equation $\frac{dA}{dt} = \lambda A - A^3$ – where A is a value characteristic of the phenomenon in question, t is a dimensionless time and λ is the control parameter – the normal form of a supercritical (continuous) pitchfork bifurcation. Figure A5.1 shows the locus of the stable equilibrium points in solid lines, and the unstable ones in dotted lines.

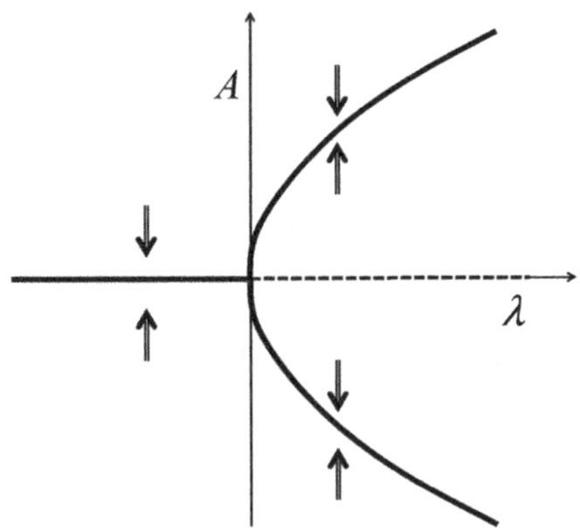

Figure A5.1. *Diagram of a supercritical pitchfork bifurcation. For the negative λ values, $A = 0$ is the only equilibrium solution. If the control parameter λ becomes positive, two stable solutions appear: $A = \pm\sqrt{\lambda}$*

The stability of the solutions is easy to study:

– we can neglect A^3 in the vicinity of $A = 0$ and obtain $dA/dt = \lambda A$, whose solution, in the form $A = A_0 e^{\lambda t}$, tends toward zero as $t \to \infty$ when $\lambda < 0$ (stability), and diverges when $\lambda > 0$ (instability);

– for $\lambda > 0$, the vicinity of the equilibrium solution $A_\pm = \pm\sqrt{\lambda}$ is studied by setting $\delta_\pm = A - A_\pm$. The evolution equation becomes $d\delta_\pm/dt + 2 A_\pm^2 \delta_\pm = 0$, whose solution $\delta_\pm = (\delta_\pm)_0 e^{-2\lambda t}$ also converges towards zero.

This study accounts for the directions of the arrows in Figure A5.1.

Appendix 6

Confined Flame

As indicated in section 3.3.2, the boundaries of inflammability of a mixture depend on the temperature and pressure, but also on the concentrations. The experiment discussed above is testament to this.

In order to gain a clearer understanding of the effects of the field of acceleration on a premixed flame, in the 1990s, experiments were carried out in parabolic flight so as to have quasi-inexistent acceleration. During these experiments, the burned gases were evacuated through the *vent line*.

In the experiments carried out in hypergravity using the centrifuge at the LCPC in Nantes, France [PRU 91b], with the same combustion chamber, the burned gases were not evacuated.

One of the difficulties encountered lies in the fact that no regulation (or draining) is possible once the centrifuge has been "started". A flame that is blown out, for instance, by the influence of the tangential acceleration, can therefore not be relit.

The absence of draining of the chamber during the experiment causes the disappearance of oxygen from the surrounding air and the accumulation of burned gases. A ground study showed that the operating points in the centrifuge were often at the boundary of stability, near to the blowout curve (Figure A6.1) which, in the case of the methane–air mixture, has the same shape as that obtained with the mixture of natural gas and air[1] (see page 407 of [VAN 68]).

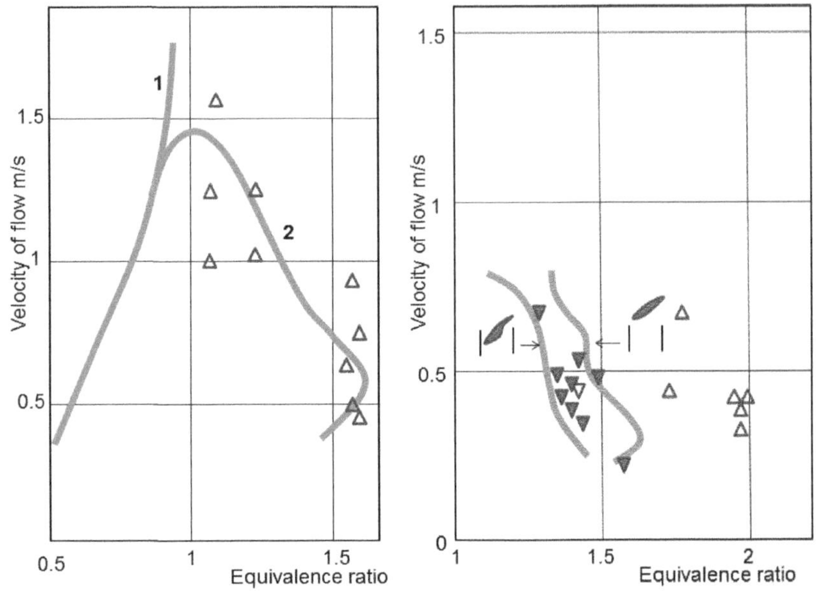

Figure A6.1. *Stability diagrams for mixtures: a) of methane and air obtained in normal gravity. 1: open casing; 2: closed casing;* Δ*: experimental points in a centrifuge, b) of propane and air (capped burner) obtained in normal gravity in a closed casing;* Δ*: experimental points in a centrifuge (non-capped burner);* ∇ *: cellular flames in parabolic flight or on the ground (non-capped burner,* $p \cong 10^5$ *Pa); ▼: cellular flames in parabolic flight or on the ground (capped burner,* $p \cong 10^5$ *Pa)*

With the propane–air mixture, the objective was to determine the domains of existence of the different structures caused by thermodiffusive instability (Figure A3.9). We can see in Figure A6.1 that all the experimental conditions of the centrifuged experiments were outside the bounds of the domain of stability. These flames did not endure; the possible influence of the acceleration when rotation is begun also needs to be taken into account.

1 The composition of the atmosphere does exert a certain amount of influence on the plot of stability of premixed Bunsen burner flames. A very clear distinction needs to be drawn between the case of an inert atmosphere and that of a generally oxidizing atmosphere, which could take part in the combustion reaction. If the atmosphere is inert, its nature also has an effect on the stall limit: a flame is less stable in an atmosphere of helium than one of nitrogen, while carbonic gas has an intermediary effect.

Appendix 7

Limits of Validity of the First-order Expansions for Diffusion Flames

A7.1. Burke–Schumann flame

Burke and Schumann studied a diffusion flame, performing an expansion into a series of Bessel functions (section 3.2.1). We pointed out earlier that the flame profiles obtained by limiting the experiment to $n = 1$ were unsatisfactory, except for the flame height, to which the authors did indeed limit themselves. To obtain the correct profile for a thin flame requires further pursuit of the expansion into a series of Bessel functions.

Figure A7.1 shows that the results obtained with a single term are incorrect, particularly on exit from the burner where the curves have to pass through $\xi = 0.5$.

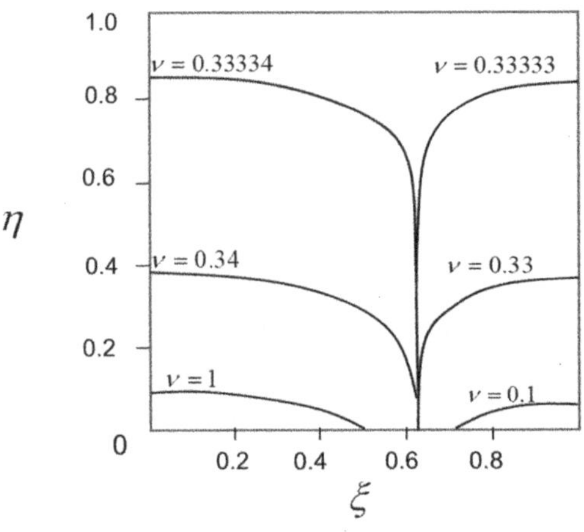

Figure A7.1. *Solutions limited to the first order of the Burke–Schumann problem. Here, C =0.5*

A7.2. Juxtaposed oxidizer/fuel flows from rectangular burners

We can show that the same is true with a series of adjacent rectangular burners. This problem can be dealt with by way of expansion into a Fourier series and 1st-order expansion gives good results only at a sufficient height above the burner.

A chamber may be constituted by a juxtaposition of oxidizer and fuel flows as indicated in Figure A7.2a. We shall make the hypothesis of rectangular conduits. If the flows of oxidizer and fuel have the same thickness a, the conditions at the openings of the burners are:

$$\begin{aligned} &(2k-2)a < x < (2k-1)\,a, \quad y=0\,(H):\ Y_H = Y_{H,0}, Y_O = Y_{O,0} = 0 \\ &(2k-1)a < x < 2ka, \quad y=0(O):\ Y_H = Y_{H,0} = 0,\ Y_O = Y_{O,0} \end{aligned} \qquad \text{[A7.1]}$$

k being the number of oxidizer and fuel phalanges. We can also choose different thicknesses for the fuel and oxidizer channels, as in Figure A7.2b, where we set $R = x_F / x_O$, which illustrates an element of the device. Using the same hypotheses as for the Burke–Schumann flame (a single velocity in the Oy direction, Shvab-Zeldovich approximation, negligible axial diffusion, stoichiometry at the level of the flame, etc.), we obtain the following expansion into a Fourier series:

$$\gamma = R(1+\nu) - \nu + \sum_{n=1}^{\infty} 2(1+\nu)\sin(n\pi R)e^{-n^2\pi^2\eta}\cos(n\pi\xi) \qquad \text{[A7.2]}$$

By setting $R = C^2$, the ratio of the normal sections and limiting ourselves to the first two terms, the thin flame profile has the equation:

$$\eta = \frac{1}{\pi^2} \ln \left[\frac{2(1+\nu)\sin(\pi C^2)}{\nu - C^2(1+\nu)} \cos(\pi \xi) \right] \qquad [A7.3]$$

Figure A7.3 gives examples of flames obtained with this simplified theory. We see certain characteristics of the predicted flames – e.g. the boundary between open and closed flames.

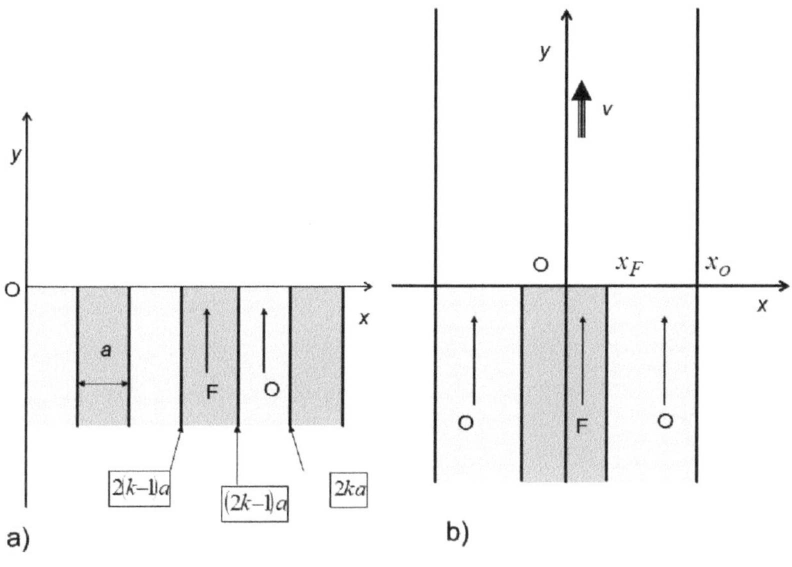

Figure A7.2. *Juxtaposed oxidizer and fuel flows: a) with equal thicknesses; b) an element of the device with unequal thicknesses and* $R = x_F / x_O \neq 0.5$

However, we can also see inadequacies. In particular, we do not see the boundary condition $\eta = 0$, except in Figure A7.3a and only then for $\nu = 1$, which gives us a vertical flame. If we examine cases where R is different to 0.5, we also note anomalies (in Figure A7.3b, ξ is not equal to R, i.e. to 0.25 in this case). Thus, necessarily, we need to take more than two terms from the Fourier series. Then, however, the expression giving the flame profile becomes more complicated, as we can see when we take only three terms.

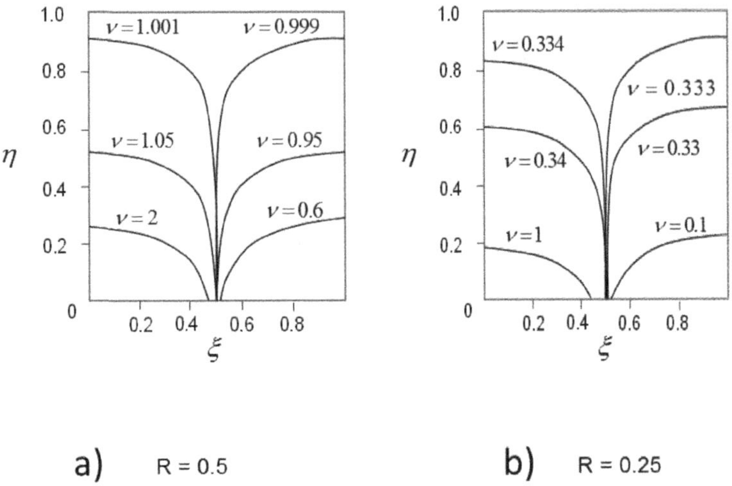

Figure A7.3. *Flame shapes obtained with two terms in the Fourier series: a)* $R = C^2 = 0.5$ *; b)* $R = 0.25$ *, so* $C = 0.5$ *; same as C in Figure A7.1*

Bibliography

[AND 86] ANDERECK C.D., LIU S.S., SWINNEY H.L., "Flow regimes in a circular Couette system with independently rotating cylinders", *Journal of Fluid Mechanics*, vol. 164, p. 166-183, 1986.

[AND 99] ANDERSSON P.-O.F., PIRJAMALI M., JÄRAS S.G., BOUTONNET-KIZLING M., "Cracking catalyst additives for sulfur removal from FFC gasoline", *Catalysis Today*, vol. 53, Issue 4, p. 517-754, 1999.

[AND 80] ANDRÉ J.-C., BARRÈRE M., Turbulence fluide (optional course), École Polytechnique, Paris, 1980.

[ARI 65] ARIS R., *Introduction to the analysis of chemical reactors*, Prentice-Hall, Englewood Cliffs, 1965.

[ARI 58] ARIS R., AMUNDSON N.R., "An Analysis of Chemical Reactor Stability and Control", Parts I-III, *Chemical Engineering Science*, 7, p. 121-155, 1958.

[ATO 69] *Atomic Energy Levels*, National Bureau of Standard, Circular 467, 1969.

[BAB 03] BABICH V., MOULIJN J.A., "Science and technology of novel processes for deep desulfurization of oil refinery streams: A review", *Fuel*, vol. 82, n° 6, p. 607-631, 2003.

[BAR 70] BARRÈRE M., PRUD'HOMME R., *Aérothermochimie des écoulements homogènes*, Gauthier-Villars, Paris, 1970.

[BAR 73] BARRÈRE M., PRUD'HOMME R., *Equations fondamentales de l'Aérothermochimie*, Masson et Cie, Paris, 1973.

[BAU 65] BAUER S., "Phenomenological theory of the relaxation phenomena in gases", *in* Warren P. Mason (*ed.*), *Physical Acoustics*, vol. II, Part A, Academic Press, New York, 1965.

[BER 84] BERGÉ P., POMEAU Y., VIDAL Ch., *L'ordre dans le chaos*, Hermann, Paris, 1984.

[BER 65] BERIATIN P., *Tables Thermodynamiques des substances*, Atomisdat, Moscow, 1965.

[BIL 55] BILOUS O., AMUNDSON N.R., "Chemical reactor stability and sensitivity", *AIChE Journal* (An Official Publication of the American Institute of Chemical Engineers.), 1, p. 513, 1955.

[BOR 95] BORGHI R., LACAS F., "Modeling of liquid-propellant spray combustion in rocket engine combustor", *Proc. 2nd Int. Symp. on Liquid Rocket Propulsion*, ONERA-Chatillon, France, 19-21 June 1995.

[BOR 00] BORGHI R., CHAMPION M., *Modélisation et théorie des flammes*, Technip, Paris, 2000.

[BOS 02] BOSKOVIC D.M., KRSTIC M., "Backstepping control of chemical tubular reactors", *Computer & Chemical Engineering*, 26, p. 1077-1085, 2002.

[BOY 05] BOYARINOV A., DUEV S., "Study of an influence of the parameters on multiplicity of steady states in the recycle systems: reactor-separating unit", *European Symposium on Computer Aided Process Engineering-15*, PUIGJANER L. and ESPUNA A. (eds.), 2005.

[BRA 12] BRAZA M., "Physique et modélisation de l'interaction fluide-structure impliquant des écoulements fortement décollés autour de corps épais à nombre de Reynolds élevé", *Journées AUM/AFM*, Rouen, 29-31 August 2012.

[BRU 70] BRUN E.A., MARTINOT-LAGARDE A., MATHIEU J., *Mécanique des fluides*, Volume III, Dunod, Paris, 1970.

[BRU 05] BRUNET S., MEY D., PEROT G., BOUCHY C., DIEHL F., "On the hydrodesulfurization of FCC gasoline: A review", *Appl. Catal. A Gen.*, vol. 278, n° 2, p. 143-172, 2005.

[BUC 96] BUCKMASTER J., "Edge-flames and their stability", *Combustion Science and Technology*, vol. 115, p. 41-68, 1996.

[BUR 48] BURGERS J.M., "A mathematical model illustrating the theory of turbulence", *Advances in Applied Mechanics*, vol. 1, p. 171-199, 1948.

[BUR 28] BURKE S.P., SCHUMANN T.E.W., MUNGAL M.G., "Diffusion flames", *First Symposium on Combustion*, Swampscott, MA, 10-14 September 1928.

[CAL 77] CALTAGIRONE J.-P., PRUD'HOMME R., "Modèle d'évolution thermique d'un matériau de forte porosité en présence d'une réaction de décomposition", *Comptes Rendus à l'Académie des Sciences*, Paris, Series B, 284, p. 491-494, 1977.

[CAN 90] CANDEL S., VEYNANTE D., LACAS F., MAISTRET E., DARABIHA N., "Coherent flame model: applications and recent extensions", in LARROUTUROU B. (ed.), *Advances in Combustion Modeling*, World Scientific, Singapore, 1990.

[CAS 70] CASTAING R., *Cours de thermodynamique statistique*, Masson et Cie, Paris, 1970.

[CHA 61] CHANDRASEKHAR S., *Hydrodynamic and hydromagnetic stability*, Clarendon, Oxford, 1961.

[CHA 01] CHASSAING P., *Turbulence en mécanique des fluides*, Cépaduès, Toulouse, 2001.

[CHO 92] CHOSSAT P., IOOSS G., *The Taylor-Couette Problem*, Springer, New York, 1992.

[CLA 64] CLARKE J.F., *The Dynamics of Real Gases*, Butterworth, London, 1964.

[CLA 78] CLAVIN P., GUYON E., "La flamme", *La Recherche*, vol. 94, November 1978.

[CLA 83] CLAVIN P., JOULIN G., "Premixed flames in large and high intensity turbulent flow", *Le Journal de Physique-Lettres*, vol. 44, 1, 1983.

[DAM 40] DAMKÖHLER G., "Der Einfluss der Turbulenz auf die Flammengeschwindigkeit in Gasgemischen", *Zeischrift für Elektrochemie*, vol. 46, p. 601-652, 1940.

[DAN 51] DANCKWERTS P.V., "Absorption by simultaneous diffusion and chemical reaction into particles of various shapes and into failing drops", *Transactions of the Faraday Society*, vol. 14, p. 1014, 1951.

[DAR 82] DARROZÈS J.-S., FRANÇOIS C., *Mécanique des fluides incompressibles*, Springer, Berlin, 1982.

[DAR 96] DARROZÈS J.-S., MONAVON A., "Analyse phénoménologique des écoulements", course at University Pierre and Marie Curie, Paris, 1996.

[DEA 70] DEADORFF J.W., "A numerical study of three-dimensional turbulent channel flow at large Reynolds number", *Journal of Fluid Mechanics*, vol. 41, p. 453-480, 1970.

[DEL 63] DELCROIX J.L, *Physique des plasmas*, Volume 1, Dunod, Paris, 1963.

[DIE 99] DIETRICH D.L., STRUK P.M., KITANO P.M., IKEGAMI M., "Combustion of interacting droplet arrays in a micro-gravity environment", *Proc. 5th International Micro-gravity Combustion Workshop*, p. 281-284, Cleveland, OH, 18-20 May 1999.

[DOM 96] DOMINGO F., VERVISCH L., "Triple flames and partially premixed combustion in autoignition of non-premixed turbulent mixtures", *Twenty-sixth Symposium (International) on Combustion*, The Combustion Institute, Pittsburgh, p. 233-240, 1996.

[DOS 08] DOS SANTOS N., Contrôle de la sélectivité des réactions d'hydrodésulfuration et d'hydrogénation en hydrotraitement : étude cinétique sur des essences craquées modèles, Doctoral thesis, Lyon 1, 10 December 2008.

[DRA 08] DRAMÉ A.K., LOBRY C., HARMAND J., RAPAPORT A., MAZENC F., "Multiple stable equilibrium profiles in tubular bioreactors", *Mathematical and Computer Modelling*, 48, 1840-1853, 2008.

[DUC 96] DUCROS F., COMTE P., LESIEUR M., "Large-eddy simulation of transition to turbulence in a boundary layer developing spatially over a flat plate", *Journal of Fluid Mechanics*, vol. 326, p. 1-36, 1996.

[DUP 86] DUPOIRIEUX F., "Calcul numérique d'écoulements turbulents réactifs et comparaison avec des résultats expérimentaux", *La Recherche Aérospatiale*, vol. 6, p. 443-453, 1986.

[DUR 87] DUROX D., PRUD'HOMME R., "Polyhedral flames of propane at low pressure", *Combustion and Flame*, vol. 70, p. 243-249, 1987.

[ECH 96] ECHEKKI T., CHEN J.H., "Unsteady strain rate and curvature effects in turbulent premixed methane/air flames", *Combustion and Flame*, vol. 106, p. 184-202, 1996.

[EIN 20] EINSTEIN A., "Schallgeschwindigkeit in teilweise dissozierten gasen", *Sitzungsbericht Preußische Akademie der Wissenschaft*", Berlin, Germany, 1920.

[ENC 08] ENCICLOPEDIA TRICCANI, 6.3 Chemical reactors, *Hydrocarbons*, vol. 5, p. 351-368, http://www.treccani.it/export/sites/default/Portale/sito/altre_aree/Tecnologia_e_Scienze_applicate/enciclopedia/inglese/inglese_vol_5/351_368_ing.pdf, 2008.

[EXO 08] EXOMER TECHNOLOGY, http://www.merichem.com/products_services/refining_technology/product_literature/pdfs/ExomerTechnicalBulletin.pdf, 2008.

[FAU 01] FAUVE S., LIBCHABER A., "Rayleigh-Bénard experiments", in BRUTER C.P. (ed.), *Bifurcation Theory, Mechanics and Physics*, Kluwer, Dordrecht, 2001.

[FAV 58] FAVRE A.J., "Equations statistiques des gaz turbulents", *Comptes Rendus à l'Académie des Sciences*, vol. 246, p. 2576, 2723, 2839, 3216, 1958.

[FAV 71] FAVRE A.J., "Statistical equations for fluctuations of temperature, entropy, concentration and vorticity in the compressible turbulent flow of a gas" (Instituto Naz. Alta Matematica, Symposia Matematica 9), Academic, New York, p. 371-390; and *Comptes Rendus à l'Académie des Sciences*, Series A, vol. 273, 1971.

[FLA 89] FLAMENT C., LE TOULLEC L., MARRAFFA L., SAGNIER P., "Inviscid nonequilibrium flow in ONERA F4 wind tunnel", *International Conference on Hypersonic Aerodynamics*, Manchester, 1989.

[FLA 90] FLAMENT C., "Chemical and vibrational non-equilibrium nozzle flow calculation by an implicit upwind method", *Proceedings of the 8th GAMM-Conference, Notes on Numerical Fluid Mechanics*, vol. 29, p. 128-137, 1990.

[FLA 93] FLAMENT C., PRUD'HOMME R., "Entropy and entropy production in thermal and chemical non-equilibrium flows", *Journal of Non-Equilibrium Thermodynamics*, vol. 18, p. 295-310, 1993.

[FRO 04] FROMENT G.F., "Modeling in the development of hydrotreatment processes", *Catalysis Today*, vol. 98, n° 1-2, p. 43-54, 2004.

[GAR 84] GARDINER JR W.C., *Combustion Chemistry*, Springer-Verlag, New York, 1984.

[GAT 01] GATIGNOL R., PRUD'HOMME R., *Mechanical and Thermodynamical Modeling of Fluid Interfaces*, Series on Advances in Mathematics for Applied Sciences, vol. 58, World Scientific, Singapore, 2001.

[GER 91] GERMANO M., PIOMELLI U., MOIN P., CABOT W.H., "A dynamic subgrid-scale eddy viscosity model", *Physics of Fluids A*, vol. 3, 7, p. 1760-1765, 1991.

[GER 97] GERMANO M., MAFFIO M., SELLO S., MARIOTTI G., "On the extension of the dynamic modelling procedure to turbulent reacting flows", *Direct and Large Eddy Simulation II*, Kluwer, Dordrecht, p. 291-300, 1997.

[GHO 95] GHOSAL S., MOIN P., "The basic equations for the large-eddy simulation of turbulent flows in complex geometry", *Journal of Computational Physics*, vol. 118, 1, p. 24-37, 1995.

[GIS 02] GISLASON J., "Phillips sulfur-removal process nears commercialization", *Oil Gas Journal*, 99, p. 74-76, 2002.

[GOD 98] GODRÈCHE C., MANNEVILLE P., *Hydrodynamics and Nonlinear Instabilities*, Cambridge University Press, Cambridge, 1998.

[GOD 53] GODSAVE G.A.E., "Studies of the combustion of drops in a fuel spray. The burning of single drops of fuel", *Fourth Symposium on Combustion*, Williams and Wilkins, Baltimore, United States, p. 818-830, 1953.

[GOL 75] GOLLUB J.P., SWINNEY H.L., "Onset of turbulence in a rotating fluid", *Physical Review Letters*, vol. 35, p. 927-930, 1975.

[GOS 96] GOSH MOULIC S., YAO L.S., "Taylor-Couette instability of traveling waves with continuous spectrum", *Journal of Fluid Mechanics*, vol. 324, p. 181-198, 1996.

[GU 03] GU X.J., EMERSON D.R., BRADLEY D., "Modes of reactions front propagation from hot spots", *Combustion and Flame*, vol. 133, p. 63-74, 2003.

[GUR 65] GURNEY J.P., NAPIER R., MONTCHILOFF I.N., "Integration of near-equilibrium flows in propulsive nozzles", *Meeting of the American Institute of Chemical Engineers*, San Francisco, 16-19 May 1965.

[GUY 91] GUYON E., HULIN J.-P., PETIT L., "Savoirs Actuels", *Hydrodynamique physique*, Inter Editions/Editions du CNRS, Paris, 1991.

[HAA 09] HAASE W., BRAZA M., REVELL A. (ed.), *DESider. A European Effort on Hybrid RANS-LES Modeling*, Notes on Numerical Fluid Mechanics and Multidisciplinary Design, vol. 103, Springer, Heidelberg, 2009.

[HAL 01] HALBERT T.R., STUNTZ G.F., BRIGNAC G.B., GREELEY J.P., ELLIS E.S., DAVIS T.J., KAMIENSKI P., MAYO S., "A Commercially Proven Technology for Low Sulfur Gasoline", *AkzoNobel Catalyst Symposium on SCANfining*, Noordwijkaan Zee, The Netherlands, 10-13 June 2001.

[HER 51] HERZBERG G., *Molecular Spectra and Molecular Structure*, vols. I and II, D. Van Nostrand, New York, 1951.

[HOL 89] HOLLANDERS R.H., MARMIGNON C., "Navier-Stokes high speed flow calculations by an implicit non-centered method", 27th *AIAA Aerospace Sciences Meeting*, Reno, 1989.

[JAN 69] JANAF THERMODYNAMICAL TABLES, The Down Chemical Company, Midland, Michigan, 1969.

[JAU 71] JAUMOTTE A.L., *Chocs et ondes de choc*, Masson, Paris, 1971.

[JEO 95] JEONG J., HUSSAIN F., "On the identification of a vortex", *Journal of Fluid Mechanics*, vol. 285, p. 69-94, 1995.

[JOM 07] JOMAAS G., LAW C.K., BECHTOLD J.K., "On transition to cellularity in expanding spherical flames", *Journal of Fluid Mechanics*, vol. 583, p. 1-26, 2007.

[JOS 76] JOSEPH D.D., *Stability of Fluid Motion*, Springer, Berlin, 1976.

[KAR 86] KARAGOZIAN A.R., MARBLE F.E., "Study of a diffusion flame in a stretched vortex", *Combustion Science and Technology*, vol. 45, 65-84, 1986.

[KAR 55] VON KARMAN TH., "Fundamental equations in aerothermochemistry", *Selected Combustion problems*, vol. II, Butterworths Scientific Publications, Agard, Liège, 1955.

[KNE 65] KNESSER H.D., "Relaxation Processes in Gases", *Physical Acoustics*, vol. II, part A, Warren P. Mason, Academic Press, New York, 1965.

[KO 02] KO Y.S., CHUNG T.M., CHUNG S.H., "Characteristics of propagating tribrachial flames in counterflow", *KSME International Journal*, vol. 16, n° 12, p. 1710-1718, 2002.

[KOR 03] KORONAKI E.D., BOUDOUVIS A.G., "Enabling stability analysis of tubular reactor models using PDE/PDAE integrators", *Computer & Chemical Engineering*, 27, p. 951-964, 2003.

[KUM 70] KUMAGAI S., SAKAI T., OKAJIMA S., "Combustion of free fuel droplets in a freely falling chamber", *13th Symposium. on Combustion*, Salt Lake City, UT, 23-29 August 1970.

[KUO 05] KUO K.K.-Y., *Principles of Combustion*, 2nd edition, Wiley, New York, 2005.

[LAM 45] LAMB H., *Hydrodynamics*, Cambridge University Press, Cambridge, 1945.

[LAN 71] LANDAU L., LIFSCHITZ E., *Mécanique des fluides*, Editions MIR, Moscow, 1971.

[LAV 89] LAVERDAN A.M., CANDEL S., "Computation of diffusion and premixed flames rolled in vortex structure", *Journal of Propulsion and Power*, vol. 5, 2, p. 134-143, 1989.

[LAW 05] LAW C.K., JOMAAS G., BECHTOLD J.K., "Cellular instabilities in expanding hydrogen propane spherical flames at elevated pressure: theory and experiment", *Proceedings of the Combustion Institute*, vol. 30, p. 159-167, 2005.

[LES 97] LESIEUR M., *Turbulence in Fluids*, Kluwer, Dordrecht, 1997.

[LIA 06] LIANG L., REITZ R. D., YI J., IYER O., "AG-equation combustion model incorporating detailed chemical kinetics for PFI/DI SI engine simulation", *Sixteenth International Engine Modeling User's Group Meeting at the SAE Congress*, Detroit, Michigan, 2 April 2006.

[LIB 80] LIBBY P.A., WILLIAMS F.A., *Turbulent Reacting Flows*, Springer, New York, 1980.

[LYO 07] LYONS K.M., "Towards an understanding of the stabilization mechanisms of lifted turbulent jet flames: experiments", *Progress in Energy and Combustion Science*, vol. 33, p. 211-231, 2007.

[MAR 77] MARBLE F.E., BROADWELL J.E., The coherent flame model for turbulent chemical reactions, Technical Report TRW-9-PU, Project Squid Headquarters, Purdue University, West Lafayette, 1977.

[MAR 85] MARBLE F.E., "Growth of a diffusion flame in the field of a vortex", in Casci C. (ed.), *Recent Advances in Aerospace Sciences*, p. 395-413, Plenum, New York, 1985.

[MAR 02] MARCHAL-GEORGE N., PICARD F., UZIO D., Process for the production of hydrocarbons with low sulfur and mercaptan content, n° US Patent 20030217951, 2002.

[MAU 13] MAURIOT Y., Etude bibliographique des approches méthodologiques pour la modélisation de la turbulence instationnaire, en vue de l'application à la stabilité BF et HF des étages de lanceurs à ergols liquides, Technical Report ONERA, N° RT 1/20318 DEFA, 2013.

[MAY 40] MAYER J.E., MAYER M.G., *Statistical Mechanics*, John Wiley and Sons, New York, 1940.

[MCG 71] MCGOWIN C.R., PERLMUTTER D.D., "Tubular reactor steady state and stability characteristics", *AIChE Journal*, 17, n° 4, p. 831-849, 1971.

[MEI 59] MEIXNER S., REIK H.G., *Thermodynamik der irreversiblen Prozesse, Handbuch der Physik* – Band III/2, Springer-Verlag, New York, 1959.

[MER 09] MERZARI E., KHAKIM A., NIKOKATA H., BAGLIETTO E., "Unsteady Reynolds-averaged Navier-Stokes: toward accurate prediction of turbulent mixing phenomena", *International Journal of Process Systems Engineering*, vol. 1, n° 1, p. 100-123, 2009.

[MIL 63] MILIKAN R.C., WHITE D.R., "Systematics of vibrational relaxation", *The Journal of Chemical Physics*, vol. 36, n° 12, p. 3209-3213, 1963.

[MOI 91] MOIN P., SQUIRES K., CABOT W., LEE S., "A dynamic subgrid-scale model for compressible turbulence and scalar transport", *Phys. Fluids A*, vol. 3, n° 11, p. 2746-2757, 1991.

[MOR 10] MORY M., *Ecoulements pour les procédés*, Hermès, Paris, 2010.

[NAJ 85] NAJA G., "Definition of an experiment on combustion of two droplets in microgravity (ST-8511)", *36th IAF Congr.*, Stockholm, Sweden, 7-12 October 1985.

[NAP 63] NAPOLITANO L.G., *Chemical Relaxation for Channel Flows of Doubly Reacting Mixtures*, Agard, London, United Kingdom, April 1963.

[NAP 64] NAPOLITANO L.G., "Unsteady wave propagation in a multi-reacting gas mixture", *15th Int. Congr. Astronautics*, Warsaw, Poland, 11 September 1964.

[NAP 66] NAPOLITANO L.G., "Transonic approximations for reacting mixtures", *Israel Journal of Technology*, vol. 4, n° 1, 1966.

[NAP 71] NAPOLITANO L.G., "Thermodynamique des systèmes composites en équilibre ou hors d'équilibre", *Mémoires des Sciences Physiques*, Fasc. 71, Gauthier-Villars, 1971.

[NAY 02] NAYAGAM V., WILLIAMS F.A., "Lewis-number effects on edge flame propagation", *Journal of Fluid Mechanics*, vol. 458, p. 219-228, 2002.

[NOC 00] NOCCA J.-L., COSYNS J., DEBUISSCHERT Q., DIDILLON B., "The domino interaction of refinery processes for gasoline quality attainment", AM-00-61 *Proceedings of the NPRA Annual Meeting*, San Antonio, United States, March 2000.

[ORL 01] ORLANDI P., *Fluid Flow Phenomena*, Kluwer, Dordrecht, 2001.

[PAP 89] PAPAMOSCHOU D., "Structure of the compressible turbulent shear layer", *27th Aerospace Science Meeting*, Reno, NV, 9-19 January 1989.

[PAR 69] PAREJA G., REILLY M.J., "Dynamic effects of recycle elements in tubular reactor systems", Parts I-III, *IEC Fundamentals*, 8, n° 4, p. 442-448, 1969.

[PAR 88] PARK C., Two-temperature interpretation of dissociation rate data for N_2 and O_2, AIAA Paper 88-0458, 1988.

[PAR 03]] PARK T.S., SUNG H.J., SUSUKI K., "Development of a nonlinear near-wall turbulence model for turbulent flow and heat transfer", *International Journal of Heat and Fluid Flow*, vol. 24, p. 29-40, 2003.

[PEL 10] PELARDY F., Impact de composés oxygénés modèles sur l'hydrodésulfuration sélective des essences issues du FCC, Doctoral thesis, Poitiers, 16 December 2010.

[PEL 82] PELCÉ P., CLAVIN P., "Influence of hydrodynamics and diffusion upon the stability limits of laminar flames", *Journal of Fluid Mechanics*, vol. 124, p. 219-237, 1982.

[PEN 57] PENNER S.S., *Chemistry Problems in Jet Propulsion*, Pergamon Press, New York, 1957.

[PET 82] PETERS N., "The premixed turbulent flame in the limit of a large activation energy", *Journal of Non-Equilibrium Thermodynamics*, vol. 7, p. 25-38, 1982.

[PET 00] PETERS N., *Turbulent Combustion*, Cambridge University Press, Cambridge, 2000.

[PLA 03] PLANTENGA F.L., LELIVELD R.G., "Sulfur in fuels: more stringent sulfur specifications for fuels are driving innovation", *Applied Catalysis A: General*, 248, p. 1-7, 2003.

[POI 92] POINSOT TH., ECHEKKI T., MUNGAL M.G., "A study of the laminar flame tip and implications for premixed turbulent combustion", *Combustion Science and Technology*, vol. 81, p. 45-55, 1992.

[POI 05] POINSOT TH., VEYNANTE D., *Theoretical and Numerical Combustion*, 2nd edition, Edwards, Flourtown, 2005.

[PRU 68] PRUD'HOMME R., "Sur l'équation de dispersion des mélanges gazeux multi réactifs", *Comptes Rendus à l'Académie des Sciences*, vol. 266, A, p. 300-301, 1968.

[PRU 69] PRUD'HOMME R., Ecoulements relaxés dans les tuyères, Publication ONERA n° 129, 1969.

[PRU 70] PRUD'HOMME R., "Analysis of transonic flow with chemical reactions by the small perturbation method. The one and two-dimensional problem", *Astronaut. Acta*, vol. 15, p. 575-586, 1970.

[PRU 88] PRUD'HOMME R., *Fluides hétérogènes et réactifs : écoulements et transferts*, Lecture Notes in Physics, vol. 304, Springer-Verlag, 1988.

[PRU 91a] PRUD'HOMME R., FLAMENT C., Entropie et production d'entropie dans les écoulements hors d'équilibre thermique et chimique, Report, Laboratoire d'aérothermique du CNRS, n° R 91-2, 1991.

[PRU 91b] PRUD'HOMME R., DUROX D., VILLERMAUX E., BAILLOT F., Influence d'un champ d'accélération sur les flammes de prémélange, Final Report of Contract n° 90/CNES/229, Laboratoire d'aérothermique du CNRS, N° RC 91-7, 1991.

[PRU 10] PRUD'HOMME R., *Flows of Reactive Fluids*, FMIA Series, vol. 94, Springer, 2010.

[PRU 12] PRUD'HOMME R., *Flows and Chemical Reactions*, ISTE, London, John Wiley and Sons, New York, 2012.

[QUI 50] QUIL F., *The Chemistry and Metallurgy of Miscellaneous Materials: Thermodynamics*, McGraw-Hill, New York, 1950.

[RAV 97] RAVET F., BAUDOUIN C., SCHULTZ J.L., "Modélisation numérique des écoulements réactifs dans les foyers de turboréacteurs", *Revue Générale de Thermique*, vol. 36, 1, p. 5-16, 1997.

[REN 00] RENARD P.H., THÈVENIN D., ROLOND J.C., CANDEL S., "Dynamics of flame-vortex interaction", *Progress in Energy and Combustion Science*, vol. 26, 3, p. 225-282, 2000.

[REY 83] REYNOLDS O., "An experimental investigation of the circumstances which determine whether the motion of water shall be direct and sinuous, and the law of resistance in parallel channels", *Philosophical Transactions of the Royal Society*, p. 51-105, 1883.

[ROC 02] ROCK K.L., "Ultra low sulfur gasoline via catalytic distillation", AIChE Spring National Meeting, 10 March 2002.

[ROS 09] ROSSINI F.D., Selected Values of Chemical Thermodynamics Properties, National Bureau of Standard, Circular 500.

[ROU 05] ROUX S., BAUDOUIN C., LARTIGUE G., POINSOT T., MEIER U., BERAT C., "Studies of mean and unsteady flow in a swirl combustor using experiments, acoustic analysis, and large eddy simulations", *Combust. Flame*, vol. 141, p. 40-54, 2005.

[SAG 98] SAGAUT P., *Introduction à la simulation des grandes échelles pour les écoulements de fluide incompressible* (Mathématiques et Applications 30), Springer-Verlag, Berlin-Heidelberg, 1998.

[SAG 01] SAGAUT P., *Large Eddy Simulation for Incompressible Flows* (Scientific Computation), Springer, Berlin, 2001.

[SAG 06] SAGAUT P., DECK S., TERRACOL M., *Multiscale and Multiresolution Approaches in Turbulence*, Imperial College Press, London, 2006.

[SAI 88] SAID R., BORGHI R., "A simulation with a 'cellular automaton' for turbulent combustion modelling", *Proc. 22nd Symposium. on Combustion*, Seattle, WA, p. 569-577, 14-19 August 1989.

[SAI 12] SAINTE-ROSE B., BERTIER N., DECK S., DUPOIRIEUX F., "Numerical simulations and physical analysis of an overexpanded reactive gas flow in a planar nozzle", *Combustion and Flame*, vol. 159, p. 2856-2871, 2012.

[SCH 85] SCHERRER D., "Combustion d'une goutte en milieu réactif avec décomposition exothermique préalable du combustible", in *Numerical Simulation of Combustion Phenomena*, Lecture Notes in Physics, Springer, Berlin, 1985.

[SCH 00] SCHLICHTING H., GERSTEN K., *Boundary-layer Theory*, 8th edition, Springer, Berlin, 2000.

[SCH 01] SCHMID P.J., HENNINGSON D.S., "Stability and transition in shear flows", *Applied Mathematical Sciences*, 142, Springer-Verlag, New York, 2001.

[SEL 04] SELLE L., LARTIGUE G., POINSOT T., KOCH R., SCHILDMACHER K.U., KREBS W., PRADE B., KAUFMANN P., VEYNANTE D., "Compressible large eddy simulation in complex geometry on unstructured meshes", *Combustion and Flame*, vol. 137, p. 489-505, 2004.

[SHA 88] SHARMA S.P., HUO W.M., PARK C., The rate parameters for coupled vibration-dissociation in a generalized SSH approximation, *15th Aerodynamic Testing Conference*, San Diego, 1988.

[SMA 63] SMAGORINSKY J., "General circulation experiment with the primitive equations", *Monthly Weather Review*, vol. 91, 3, p. 99-164, 1963.

[SON 03] SONG C., "An overview of new approaches to deep desulfurization for ultra-clean gasoline, diesel fuel and jet fuel", *Catalysis Today*, vol. 86, n°1-4, p. 211-263, 2003.

[SPA 53] SPALDING D.B., "Studies of the combustion of drops in a fuel spray", in Combustion Institute (ed.), *Fourth Symposium on Combustion*, Williams and Wilkins, Baltimore, p. 847-864, 1953.

[SPA 73] SPALDING D.B., Turbulence models and their experimental verification: turbulence effects on combustion, Imperial College of Science and Technology, London, 1973.

[SUL 03] SULLIVAN P.P., HORST T.W., LENSCHOW D.H., MOENG C.H., WEIL J.C., "Structure of subfilter-scale fluxes in the atmospheric surface layer with application to large eddy simulation modelings", *Journal of Fluid Mechanics*, vol. 482, p. 101-139, 2003.

[TAY 23] TAYLOR G.I., "Stability of a viscous liquid contained between two rotating cylinders", *Philosophical Transactions of the Royal Society*, vol. A223, p. 289-343, 1923.

[TØP 96] TØPSOE H., CLAUSEN B.S., MASSOTH F.E., *Hydrotreating Catalysis*, Science and Technology, Springer Verlag, vol. 11, 1996.

[TRA 04] TRAMBOUZE P., EUZEN J.P., *Chemical Reactors: from Design to Operation*, 2nd edition, Technip, New York, 2004.

[TRI 88] TRITTON D.J., *Physical Fluid Dynamics*, 2nd edition, Clarendon Press, Berlin, 1988.

[TRO 94] TROUVE A., POINSOT TH., "The evolution equation for the flame surface density in turbulent premixed combustion", *Journal of Fluid Mechanics*, vol. 278, p. 1-31, 1994.

[VAN 68] VANTIGGELEN A *et al.*, *Oxydations et combustions*, Volume 1, Technip, Paris, 1968.

[VEY 97] VEYNANTE D., TROUVE A., BRAY K.N.C., MANTEL T., "Gradient and counter gradient scalar transport in turbulent premixed flames", *Journal of Fluid Mechanics*, vol. 332, p. 263-293, 1997.

[VIL 82] VILLERMAUX J., *Génie de la réaction chimique*, Lavoisier, Paris, 1982.

[VIN 86] VINCENTI W.G., *Introduction to Physical Gas Dynamics*, John Wiley and Sons, New York, 1986.

[WAU 94] WAUQUIER J.P., *Le raffinage du pétrole* : Pétrole brut, Produits pétroliers, Schémas de fabrication, Technip, Paris, 1994.

[WEG 65] WEGENER P.P., CHU B.T., KLIKOFF W.A., "Weak waves in relaxing flow", *Journal of Fluid Mechanics*, vol. 23, part 4, 1965.

[WIL 70] WILLIAMS F.A., "An approach to turbulent flame theory", *Journal of Fluid Mechanics*, vol. 40, p. 401-421, 1970.

[WIL 75] WILLIAMS F.A., "Recent advances in theoretical descriptions of turbulent diffusion flames", in Murthy S.N.B. (ed.), *Turbulent Mixing in Nonreactive and Reactive Flows*, p. 189-208, Plenum, New York, 1975.

[WIL 85] WILLIAMS F.A., *Combustion Theory*, 2nd edition, Benjamin Cummings, Menlo Park, CA, 1985.

[WOO 75] WOODS L.C., *The Thermodynamics of Fluid Systems*, Clarendon Press, Oxford, 1975.

[YEE 85] YEE H.C., HARTEN A., "Implicit TVD schemes for hyperbolic conservation laws in curvilinear coordinates", *Proc. of the AIAA 7th Computational Fluid Dynamics Conference*, Cincinnati, 1985.

[YEE 87] YEE H.C., "Upwind and symmetric shock-capturing schemes", NASA TM 89464, 1987.

[ZEL 80] ZELDOVICH Y.B., "Regime classification of an exothermic reaction with non uniform initial conditions", *Combustion and Flame*, vol. 39, p. 211-214, 1980.

Index

P

R

S

Printed and bound by CPI Group (UK) Ltd, Croydon, CR0 4YY

27/09/2026

14970452-0001